ASSESSING THE NATIONAL STREAMFLOW INFORMATION PROGRAM

Committee on Review of the USGS
National Streamflow Information Program

Water Science and Technology Board
Division on Earth and Life Studies

NATIONAL RESEARCH COUNCIL
OF THE NATIONAL ACADEMIES

THE NATIONAL ACADEMIES PRESS
Washington, D.C.
www.nap.edu

THE NATIONAL ACADEMIES PRESS   500 Fifth Street, N.W.   Washington, DC 20001

NOTICE: The project that is the subject of this report was approved by the Governing Board of the National Research Council, whose members are drawn from the councils of the National Academy of Sciences, the National Academy of Engineering, and the Institute of Medicine. The members of the committee responsible for the report were chosen for their special competences and with regard for appropriate balance.

This study was supported by Cooperative Agreement No. 01HQAG0030 between the National Academy of Sciences, U.S. Department of the Interior, and U.S. Geological Survey. Any opinions, findings, conclusions, or recommendations expressed in this publication are those of the author(s) and do not necessarily reflect the views of the organizations or agencies that provided support for the project.

International Standard Book Number 0-309-09210-8 (Book)
International Standard Book Number 0-309-53197-7 (PDF)
Library of Congress Control Number 2004110889

Assessing the National Streamflow Information Program is available from National Academies Press, 500 Fifth Street, N.W., Lockbox 285, Washington, DC 20055; (800) 624-6242 or (202) 334-3313 (in the Washington metropolitan area); Internet, http://www.nap.edu

Cover design by Michele De La Menardiere, National Academy Press. Photo of the Quinault River on the Olympic Peninsula, Washington, courtesy of Jim O'Connor, U.S. Geological Survey. Price Lake image courtesy of North Carolina Department of Commerce, Division of Tourism, Film, and Sports Development, http://www.visitnc.com/media/ media_images.asp.

Copyright 2004 by the National Academy of Sciences. All rights reserved.

Printed in the United States of America

THE NATIONAL ACADEMIES
Advisers to the Nation on Science, Engineering, and Medicine

The **National Academy of Sciences** is a private, nonprofit, self-perpetuating society of distinguished scholars engaged in scientific and engineering research, dedicated to the furtherance of science and technology and to their use for the general welfare. Upon the authority of the charter granted to it by the Congress in 1863, the Academy has a mandate that requires it to advise the federal government on scientific and technical matters. Dr. Bruce M. Alberts is president of the National Academy of Sciences.

The **National Academy of Engineering** was established in 1964, under the charter of the National Academy of Sciences, as a parallel organization of outstanding engineers. It is autonomous in its administration and in the selection of its members, sharing with the National Academy of Sciences the responsibility for advising the federal government. The National Academy of Engineering also sponsors engineering programs aimed at meeting national needs, encourages education and research, and recognizes the superior achievements of engineers. Dr. Wm. A. Wulf is president of the National Academy of Engineering.

The **Institute of Medicine** was established in 1970 by the National Academy of Sciences to secure the services of eminent members of appropriate professions in the examination of policy matters pertaining to the health of the public. The Institute acts under the responsibility given to the National Academy of Sciences by its congressional charter to be an adviser to the federal government and, upon its own initiative, to identify issues of medical care, research, and education. Dr. Harvey V. Fineberg is president of the Institute of Medicine.

The **National Research Council** was organized by the National Academy of Sciences in 1916 to associate the broad community of science and technology with the Academy's purposes of furthering knowledge and advising the federal government. Functioning in accordance with general policies determined by the Academy, the Council has become the principal operating agency of both the National Academy of Sciences and the National Academy of Engineering in providing services to the government, the public, and the scientific and engineering communities. The Council is administered jointly by both Academies and the Institute of Medicine. Dr. Bruce M. Alberts and Dr. Wm. A. Wulf are chair and vice chair, respectively, of the National Research Council.

www.national-academies.org

COMMITTEE ON REVIEW OF THE USGS NATIONAL STREAMFLOW INFORMATION PROGRAM

DAVID R. MAIDMENT, *Chair,* The University of Texas, Austin
A. ALLEN BRADLEY, JR., University of Iowa, Iowa City
BENEDYKT DZIEGIELEWSKI, Southern Illinois University, Carbondale
    (through December 31, 2002)
RICHARD HOWITT, University of California, Davis
N. LEROY POFF, Colorado State University, Fort Collins
KAREN L. PRESTEGAARD, University of Maryland, College Park
STUART S. SCHWARTZ, Cleveland State University, Cleveland
DONALD I. SIEGEL, Syracuse University, Syracuse, New York
MARY W. STOERTZ, Ohio University, Athens
DAVID G. TARBOTON, Utah State University, Logan
KAY D. THOMPSON, Washington University, St. Louis, Missouri

Staff

WILLIAM S. LOGAN, Project Director
ANITA A. HALL, Program Associate
DIONNE PRINGLE, Anderson Intern

Editor

FLORENCE POILLON

WATER SCIENCE AND TECHNOLOGY BOARD

RICHARD G. LUTHY, *Chair,* Stanford University, Stanford, California
JOAN B. ROSE, *Vice Chair,* Michigan State University, East Lansing
RICHELLE M. ALLEN-KING, State University of New York at Buffalo
GREGORY B. BAECHER, University of Maryland, College Park
KENNETH R. BRADBURY, Wisconsin Geological and Natural History Survey, Madison
JAMES CROOK, Water Reuse Consultant, Norwell, Massachusetts
EFI FOUFOULA-GEORGIOU, University of Minnesota, Minneapolis
PETER GLEICK, Pacific Institute for Studies in Development, Environment, and Security, Oakland, California
JOHN LETEY, JR., University of California, Riverside
CHRISTINE L. MOE, Emory University, Atlanta, Georgia
ROBERT PERCIASEPE, National Audubon Society, Washington, D.C.
JERALD L. SCHNOOR, University of Iowa, Iowa City
LEONARD SHABMAN, Virginia Polytechnic Institute and State University, Blacksburg
R. RHODES TRUSSELL, Trussell Technologies, Inc., Pasadena, California
KARL K. TUREKIAN, Yale University, New Haven, Connecticut
HAME M. WATT, Independent Consultant, Washington, D.C.
JAMES L. WESCOAT, JR., University of Illinois at Urbana-Champaign

Staff

STEPHEN D. PARKER, Director
LAURA J. EHLERS, Senior Program Officer
JEFFREY W. JACOBS, Senior Program Officer
WILLIAM S. LOGAN, Senior Program Officer
LAUREN E. ALEXANDER, Program Officer
MARK C. GIBSON, Program Officer
STEPHANIE E. JOHNSON, Program Officer
M. JEANNE AQUILINO, Financial and Administrative Associate
ELLEN A. DE GUZMAN, Research Associate
PATRICIA JONES KERSHAW, Study/Research Associate
ANITA A. HALL, Program Associate
DOROTHY K. WEIR, Senior Program Assistant

Preface

This report is a product of the Committee on Review of the USGS National Streamflow Information Program. This committee was formed in response to discussions held between the U.S. Geological Survey (USGS) and the National Research Council (NRC) Committee on USGS Water Resources Research. The committee works under the auspices of the Water Science and Technology Board of the National Research Council.

Streamflow data and information is an aspect of water science that profoundly affects people's lives. Flood forecasting and drought management; water supply for agriculture, industry, and cities and towns; maintaining instream flows for game fish and other aquatic species and for canoeing and kayaking; and enforcing legal agreements between states and nations—all depend on the availability of high-quality information about the water elevation and discharge of our rivers and streams.

The U.S. Geological Survey is the primary federal agency charged with acquisition and quality control of raw data and its transformation into usable information. Users range from local consultants and municipalities to whitewater rafters, and from academic institutions to federal agencies such as the U.S. Army Corps of Engineers (USACE) and members of Congress. The water resources discipline of the USGS has more than a century of experience in streamgaging. However, societal needs change, science and technology move forward, and the USGS has evolved as well. For example, satellite data transmission, Doppler radar for precipitation estimates, and improvements in flood forecast models have combined to make USGS streamflow data much more valuable for flood forecasting than in the past.

This report concerns the National Streamflow Information Program (NSIP). The NSIP itself was proposed by the USGS to Congress in 1999. Although the gages that comprise it are not new—some of them have been around for half a century or more—the concept of a network of gages, other kinds of data sources, and integrated research designed to meet *national* needs is new. The USGS therefore asked the NRC to provide feedback on the nascent program.

The committee heard the first presentations on this topic in October 2001. During the next 24 months, the committee met with numerous experts from within and outside the USGS. We are particularly grateful for the assistance of Edmund D. (Ned) Andrews (USGS), Gregor T. Auble (USGS), Jerad D. Bales (USGS), Thomas R. Carroll (National Weather Service), John E. Costa (USGS), Robert M. Hirsch (USGS), Robert B. Jacobson (USGS), Joseph L. Jones (USGS), Matthew C. Larsen (USGS), Daniel R. Luna (National Weather Service), Gail E. Mallard (USGS), Ronald C. Mason (USACE), Gary P. McDevitt (National Weather Service), J. Michael Norris (USGS), Jim E. O'Connor (USGS), Harold H. Opitz (National Weather Service), and J. Dungan Smith (USGS). Committee members then drafted individual contributions and deliberated as a group to achieve consensus on the content of this report.

This report has been reviewed in draft form by individuals chosen for their diverse perspectives and technical expertise, in accordance with procedures approved by the NRC's Report Review Committee. The purpose of this independent review is to provide candid and critical comments that will assist the NRC in making its published report as sound as possible and will ensure that the report meets institutional standards for objectivity, evidence, and responsiveness to the study charge. The review comments and draft manuscript remain confidential to protect the integrity of the deliberative process.

We wish to thank the following individuals for their review of this report:

J. David Allan, University of Michigan
Roger C. Bales, University of California, Merced
Lawrence E. Band, University of North Carolina-Chapel Hill
Kaye Brubaker, University of Maryland
Emery T. Cleaves, Maryland Geological Survey
Katherine K. Hirschboeck, University of Arizona
Marc Ribaudo, U.S. Department of Agriculture-Economic Research Service

Although the reviewers listed above have provided many constructive comments and suggestions, they were not asked to endorse the conclusions or recommendations, nor did they see the final draft of the report before its release. The review of this report was overseen by Dr. M. Gordon "Reds"

Preface

Wolman, of Johns Hopkins University. Appointed by the National Research Council, Dr. Wolman was responsible for making certain that an independent examination of this report was carried out in accordance with institutional procedures and that all review comments were carefully considered. Responsibility for the final content of this report rests entirely with the authoring committee and the institution.

This committee is not the first to comment on the NSIP program and will likely not be the last. We do hope that some of the ideas generated in this report will stimulate further discussions that must take place, not only within the USGS, but also with congressional staff, state and federal agencies, and other generators and users of streamflow data and information. We trust that these discussions will lead to new and better ways to integrate this information into the human and natural world.

David R. Maidment, *Chair*
Committee on Review of the USGS
National Streamflow Information Program

Contents

EXECUTIVE SUMMARY 1

1 INTRODUCTION 13
    The National Streamflow Information Program, 15
    Statement of Task, 17
    Organization and Content of this Report, 18

2 GAGING THE NATION'S STREAMS 19
    A History of the Study of Rivers at the USGS, 19
    What Is a Gaging Site, 26
    The NSIP Gaging Network, 32
    Role of Other Agencies in Supporting Streamgaging, 32
    Streamflow Network Design in Other Countries, 33
    Value of a National Streamflow Information Program, 39
    Rationale for Federal Support, 44
    Summary, 46

3 SELECTION OF NSIP BASE GAGE LOCATIONS 47
    The Five Criteria for Siting NSIP Streamgages, 49
    Assessment of the Distribution of Gage Site Locations, 61
    Summary, 66

4 STREAMFLOW NETWORK DESIGN 68
    Statistical Models, 69
    Coverage Models, 78
    The NSIP Network as a Coverage Model, 83
    Recommendations of the Interstate Council on Water Policy, 84

Network Design Goals: Contrasting NSIP with State-Designed Streamflow Networks, 89
NSIP Network Design: From Data to Information, 92
Summary, 98

5   STREAMFLOW INFORMATION   100
Intense Data Collection During Floods and Droughts, 101
Regional and National Streamflow Assessments, 106
Enhanced Information Delivery, 109
Methods Development and Research, 114
Summary, 118

6   CONTRIBUTIONS OF NSIP TO RIVER SCIENCE   120
River Science Opportunities Created by the NSIP, 121
Information Needs for River Science, 130
Summary, 134

7   SUMMARY AND CONCLUSIONS   135
Rationale for Federal Support of a National Network, 136
The Base Gage Network, 137
Other NSIP Components, 142
Adaptive Management, 144
River Science, 145

REFERENCES   146

APPENDIX A

Biographical Sketches of Members of the Committee on Review of the USGS National Streamflow Information Program   161

Executive Summary

The measurement of streamflow is part of the documentation of the history of the nation. Knowledge of the flow of water in the nation's streams and rivers plays a vital role in flood protection, water supply, pollution control, and environmental management. In 1998, at the request of Congress, the U.S. Geological Survey (USGS) prepared a report entitled *A New Evaluation of the USGS Streamgaging Network* (USGS, 1998), stating that the network's ability to meet long-standing federal goals had declined because of an absolute loss of streamgages, a disproportionate loss of streamgages with a long period of record, and the declining ability of the USGS to continue operating high-priority streamgages when partners discontinue funding. That report also stated that new resource management issues and data delivery capabilities have increased the demand for streamflow information and that new technologies and methodologies have to be developed to improve the reliability of streamflow information and decrease its cost. Most importantly, this report proposed creation of the National Streamflow Information Program (NSIP).

As part of NSIP's design, the USGS established five goals to satisfy minimum national streamflow information needs (Box ES-1), and conducted an analysis to locate gage sites that meet these goals; these sites constitute NSIP's "base streamgage network." About 70 percent of the sites selected are already gaged. The USGS intends to support these gages entirely with federal funds and, in future years, to increase the percentage of these sites that are gaged. In addition to enhanced streamgaging, the USGS envisages four other components of NSIP Box ES-1). The USGS asked the National Research Council to review this proposed program (Box ES-2).

> **BOX ES-1**
> **Goals and Components of the**
> **National Streamflow Information Program**
>
> The five components of NSIP are the following:
>
> 1. An enhanced nationwide base streamgage network that would be 100 percent federally funded
> 2. Intense data collection during floods and droughts, and additional analysis of these data
> 3. Periodic regional and national assessments of streamflow characteristics
> 4. Enhanced information delivery
> 5. Methods development and research
>
> The first of these components, the base streamgage network, is designed to meet the following goals:
>
> 1. Meet legal and treaty obligations on interstate and international waters
> 2. Support flow forecasting
> 3. Measure river basin outflows for calculating regional water balances over the nation
> 4. Monitor sentinel watersheds for long-term trends in natural flows
> 5. Measure flow for water quality needs

FEDERAL SUPPORT FOR A BASE STREAMGAGE NETWORK

Independent of the USGS's long experience in providing consistent, quality-assured streamflow data, a national streamflow information program merits federal support because streamflow information supports national interests (e.g., interstate water supply disputes) in addition to local or private interests. In fact, streamflow information has many of the properties of a public good, because everyone benefits, whether they pay or not, and benefits to additional "users" come at no additional cost. The public also values efficiency and equality of access, both of which are characteristics of this federally provided information.

The national economy is inseparably bound to the adequacy of water supplies. By mass, consumptive use of water is the single largest material flow in the U.S. economy by a factor of more than 20. The national interest in economic information on commodity flows has long been recognized and supported with federal funding. Unfortunately, much of the funding for the network comes from cost-sharing partnerships between the USGS, other federal agencies, and

> **BOX ES-2**
> **Statement of Task for the**
> **Committee on Review of the**
> **USGS National Streamflow Information Program**
>
> The nation requires streamflow information for a variety of purposes to address local, regional, and national water-management issues. The USGS has developed a conceptual plan, the National Streamflow Information Program (NSIP), as a new approach to the acquisition and delivery of streamflow information. NSIP as proposed by the USGS includes a set of minimum national streamflow information needs that should be met by the network and design characteristics of the program.
>
> Streamflow information also supports analysis of river science, including interaction of hydrology, geomorphology and ecology. For example, changes in land use, climate change, reservoir construction, and other factors, cause changes in streamflow through time.
>
> Therefore, the committee will evaluate the program with respect to:
>
> • The minimum national streamflow information needs that should be met by the network, including those related to interstate and international waters, flow forecasts, river basin outflows, sentinel watersheds, and water quality.
> • The components of the NSIP plan that are reasonable, appropriate, and sufficient, including an enhanced nationwide streamgaging network with a larger share of national funding, intensive data collection during major floods and droughts, periodic regional and national assessments of streamflow characteristics, enhanced streamflow information delivery to customers, and methods development and research.
> • The ways a National Streamflow Information Program should support the data and information needs of the various fields of river science, in addition to meeting its operational objectives.

state and local water resource agencies. Reliance on these agencies to support the nation's needs for streamflow information has caused troubling instability and declines in the network.

Federal support of a base streamgaging network is recommended to ensure the long-term viability of this network for national needs and is justified because many national interests are served by providing streamflow information, which has many properties of a "public good."

NSIP COMPONENT 1: THE BASE GAGE NETWORK

Program Goals

In addition to the five goals for the NSIP base gage network recommended by the USGS (Box ES-1), many others could be formulated. Indeed, the Streamgaging Task Force of the Department of Interior's Advisory Committee on Water Information discussed nine additional possible goals relating to National Flood Insurance Program communities, National Pollution Discharge Elimination System (NPDES) permits, canoeists and rafters, federal lands, reservoirs, migratory fish, navigation, and others (ICWP, 2002). However, the goals selected by the USGS are important nationally and were chosen well. The committee does not concur with the recommendation of the Interstate Council on Water Policy (ICWP) to delete the NSIP sites measuring flows at many state border sites.

Goal 2, originally conceived as supporting only National Weather Service (NWS) forecast points, should be broadened to also supporting Natural Resource Conservation Service (NRCS) forecast sites. USGS streamgages are clearly indispensable in the NWS's streamflow predictions, and this goal already dictates more than 60 percent of the sites selected for NSIP gages. However, the NRCS also forecasts flows—for water supply, drought management and response, hydroelectric and thermal power production, irrigation, and navigation in western states. These forecast sites should be added to the NSIP base gage network. A greater degree of collaboration between the USGS, NWS, and NRCS in planning and locating future gage sites and forecast points would be beneficial, especially in arid and semiarid states with growing populations, where intermittent or "flashy" streamflow creates forecasting challenges.

The five NSIP goals reflect areas of compelling national interest in streamflow information and are an appropriate foundation for the National Streamflow Information Program. The set of minimum national streamflow information needs that underlie these goals are reasonable and appropriate. The national distribution of NSIP base gage sites is also reasonable.

Streamgage Network Design Methods

The question of where to sample (gage) streamflow to estimate streamflow at any point of interest is a central one for hydrologic data collection agencies worldwide. One traditional approach relies on statistical methods, including correlation of flows at pairs of gages, regression analyses to estimate flow characteristics, entropy analysis, and other approaches. Statistical methods can be effective for assessing the information content of existing streamgage records for limited goals. However, the nation's need for streamflow information is extraordinarily diverse and dynamic, and is not readily reduced to a small number of information metrics. Further, these methods assume hydrologic homogeneity throughout the study area—a poor assumption for a national network.

The approach the USGS has taken with NSIP is to generate a "coverage" model in which a goal or set of goals is established (Box ES-1) and a set of gage sites is selected using a performance metric to evaluate national coverage for each of these goals. The coverage model treats gage network design as a facility location problem—an approach that has not been widely used in the design of streamgage networks. However, coverage models are used in many other fields, such as locating fire stations so that each household is within five minutes of a fire station. The addition or omission of a particular goal corresponds to the addition or omission of a specific set of gage site locations. The coverage method provides "yes or no" answers about where gage sites should be located, regardless of whether gaging has been done there previously.

It is conceptually appealing to think of streamgage network design by assigning a "value" to individual gages and then optimizing the value or utility of the entire network. However, the list of factors that could be used to value information for individual gages is long and diverse and incorporates inherently noncommensurate uses (e.g., real-time reservoir operation and scientific evaluation of regional water balances). The value of information cannot be quantified independently from the decision-making process in which that information will be used. Moreover, weighting and combining estimated information value (e.g., property loss versus loss of life) inherently engenders local and regional (not just national) preferences and values.

Within the larger context of coverage, methods such as statistical modeling based on hydrologic regionalization and estimating the value of individual gages can be useful. They can guide incremental decisions to add or eliminate individual gages within a local or regional network serving narrow, well-defined goals by, for example, ranking gages in order of their marginal regional information content. In contrast, the breadth of both the national goals and the hydroclimatic variation spanned by the NSIP network cannot meaningfully be reduced to a concise set of valuation measures. Therefore, the most appropriate role of these methods for NSIP is supporting the analysis of incremental refinements to local and regional hydrologic networks within the broader context of the NSIP network design.

The method of designing the NSIP base gage network by establishing national goals and then using geographic information system (GIS) based methods to select sites to provide national coverage of these goals is reasonable. This is an effective use of geospatial information and analysis, and provides an innovative new method for streamgage network design. Statistical methods should only be used to justify incremental decisions to add or eliminate individual gages within a local gage network serving narrow, well-defined goals.

NSIP COMPONENT 2: INTENSE DATA COLLECTION DURING FLOODS AND DROUGHTS

The opportunistic collection of hydrologic, climatologic, geomorphic, and biological data during extreme events is a high-return, cost-effective activity, well suited to both the mission and the expertise of USGS. The value of this information is high, especially when the protocols are integrated with continuous improvement in techniques. Such protocols and measurement techniques have to address the unique challenges in monitoring flow extremes, which are usually outside the range of direct measurements used to establish flow rating curves and are often poorly measured by conventional techniques. The USGS's outstanding studies of the 1993 Mississippi River provide compelling models for integrated interdisciplinary study of extreme events, integrating expertise in water resources, channel and floodplain morphology, sediment transport, hyporheic processes, and ecosystem response.

Such data collection and analysis is a strength of the USGS and should be continued.

NSIP COMPONENT 3: PERIODIC REGIONAL AND NATIONAL ASSESSMENTS OF STREAMFLOW CHARACTERISTICS

The essential role of water in the U.S. economy, along with growing demands and policy conflicts, creates a vital national interest in consistent, objective, regional and national assessments of the nation's water resources. The information content of the whole streamflow network is more than the sum of the parts. Regional and national assessment is an integrating lens that can bring a sharpened focus to data collection and information generation within NSIP. However, the national interest is not well served

Executive Summary 7

by the current paradigm in which interpretive studies are supported mainly by cooperators. Assessment raises significant political and scientific challenges, including evaluating risks to and reliability of the nation's water resources; accounting for irrigation return flows, in-stream uses, and surface water-groundwater interactions; integrating economic and social dimensions of water use; projecting hydroclimatic variability; and addressing a host of issues at the intersection of natural, engineered, and human systems.

Such studies are fundamental to NSIP and should be continued.

NSIP COMPONENT 4: ENHANCED INFORMATION DELIVERY

The revolution in information technology is changing old paradigms regarding the access, storage, and generation of information—including streamflow information. The steady increase in data telemetry and near-real-time data delivery on the Internet has vastly expanded the awareness and utility of national streamgaging data. The USGS is committed to reengineering its data delivery paradigms, despite the variability in funding for core programs. Users are unequivocally enthusiastic about the new modes of information delivery. Notwithstanding these successes, even richer opportunities exist to enhance the content, value, and national benefits gained from streamflow information.

Streamflow information that most users see consists of tabular discharge measurements derived from "unit values," that is, stage measurements at points in space and time, for example, each 15 minutes. Presently, the USGS displays these unit values on the Internet as part of its real-time streamflow information system, but limits the publication of historical streamflow data to daily values. Significant information content is lost in this process, particularly for studies of floods on small watersheds, where the whole flood may come and go within a few hours. Publication of unit value data for the historical streamflow records would be a significant information delivery enhancement.

Streamflow data can support a far richer interpretation if combined with geospatial information, as indeed the USGS has done in designing the NSIP base gage network. Streamgaging points can be associated with information about the stream network (e.g., network topology codified in the National Hydrography Dataset) and its associated subbasin, the geomorphic and hydraulic features of the stream channel, floodplain characteristics, and other landscape attributes.

Since a streamgaging site has a subwatershed just upstream of it, whose drainage passes through that gage site before reaching any other site, the selection of a set of gage site locations can be associated with a map or dataset of subwatersheds draining to those sites. The effect of associating a dataset of subwatersheds with a set of selected gage sites is to more intimately connect the land and water systems of the nation. This is important because it provides a mechanism for using geospatial information to generalize measured streamflow to ungaged locations where information is needed, such as at the boundaries of Total Maximum Daily Load (TMDL) segments for water quality management or at the upstream ends of reaches for Federal Emergency Management Agency floodplain map delineation. In this manner, streamflow measurement and associated geospatial interpolation of flows can support streamflow information estimates at any location on the stream network of the nation. At present, through its Streamstats program, the USGS is developing the technology to support estimating streamflow statistics at any location on the stream network. This approach also might usefully be applied to geospatial interpolation of flow records to ungaged locations where streamflow information is desired.

Supporting this rich interpretation requires an enhanced data delivery system capable of handling a diverse data mixture including tabular data, geospatial data (GIS layers), remotely sensed images, and multidimensional data fields such as stream velocities. An enhanced data delivery system should also explore emerging modes of data delivery, such as direct satellite delivery and radio-frequency "push" technology, to transmit streaming information (e.g., for the NWS). Users could then tailor this information to their needs (e.g., streamflow characteristics at ungaged points, estimating channel characteristics for Hydrologic Engineering Center (HEC) modeling, flood inundation simulations).

Enhanced data delivery is an important and highly valued component of NSIP. The USGS should provide access to a broader range of geospatially linked data (unit values, channel cross sections, remotely sensed images, velocity fields, stream network position, and landscape attributes) to enable richer data interpretation than is presently done.

NSIP COMPONENT 5: METHODS DEVELOPMENT AND RESEARCH

The USGS is investigating new methods for measurement of streamflow and water quality. These include the use of radar for surface water

Executive Summary 9

velocity and water depth measurement, and the deployment of acoustic Doppler profilers for measurement of the cross-sectional distribution of velocity. The intent is to create the "gaging station of the future" wherein the measurements of flow, cross-sectional bed profile, and velocity are accomplished and recorded continuously. Another goal is the rapid reconnaissance of flow at ungaged locations during floods.

Water quality parameters such as conductivity, temperature, pH, dissolved oxygen, turbidity, and total chlorophyll are being sensed continuously and connected by regression equations to provide estimates of nutrients, bacteria, and other constituents of concern continuously through time. This effort places water quality measurement in the same mode as streamflow measurement and is a very significant enhancement over spot sampling of water quality constituents at periodic intervals. For example, for TMDL studies, this approach may illuminate under what flow conditions and at what times the water quality standards supporting reasonable water use are not being met and, thus, provide guidance for closer attainment of these standards.

Advances in techniques for remote sensing and analysis of information are accelerating. Besides the techniques just mentioned, video image analysis and new forms of LIDAR (light detection and ranging) show promise of providing significant improvements in streamflow and streambed measurement.

With due care in ensuring comparability of data produced by traditional streamgaging methods and new technologies, the USGS is encouraged to aggressively pursue new technologies for streamflow and water quality measurement with a view to accelerating the implementation of time- and labor-saving flow measurement techniques and continuous water quality monitoring, as soon as practicable.

In addition to evaluating the goals and components of the NSIP, several broader issues related to the program were examined. These included the overall role of the NSIP as an information program, the integration of the principles of adaptive management into such a program, and the potential for the NSIP to contribute to the USGS's future work in "river science" (i.e., integrated research involving all of the major disciplines at the USGS).

NSIP AS AN INFORMATION NETWORK

The historical specialization of the streamgage program has fostered a cultural separation of data collection and data use. Conceived and structured as a national information program, the NSIP embodies the broader vision required to meet the nation's needs in the twenty-first century. It is important to highlight the difference between the need for data from data collection points where streamflow or some other property is measured, and the need for information, with corresponding information points for which streamflow information is desired and generated from available data. The locations identified as sites for NSIP gages represent a well-defined set of locations where streamflow information would clearly support the five goals. However, locating a permanent streamflow gaging station at every point is not necessarily the best way to meet the information needs. In many applications the need for information may be satisfied with intermittent or remotely sensed measurements or with regionalized analytical approaches that do not require direct measurement. The benefit of such approaches will be realized as expanding populations identify new locations at which streamflow information is needed. This goal is valued by the public and is an appropriate task within the scope of the NSIP.

The ultimate goal of the NSIP should be to develop the ability to generate streamflow information (with quantitative confidence limits) at any location, gaged or ungaged, on the stream network.

ADAPTIVE MANAGEMENT

Although the five goals (Box ES-1) reflect compelling areas of national interest, the USGS's role as the nation's source of unbiased streamflow information creates unique streamflow information demands. In contrast to an individual user whose streamflow information needs are driven by well-defined operational needs (e.g., hydropower production, flood warning), the USGS has the added responsibility to develop streamflow information to satisfy the future needs of the nation. For example, the Hydro-Climatic Data Network, which allows national analysis of the trends in streamflow (the integrator of climate, topography, geology, and land use), is a "discovered" streamgage network, serendipitously maintained within the national network through the cumulative effect of unrelated decisions to maintain gaging at these sites. When gaging was initiated at many of the sites with 50- to 100-year records, detection of trends in climate change was an unimagined use of streamgage data.

The USGS's role requires forethought to provide the basic data collection and information that the nation will need decades from now. These needs, which may be most valuable to future users of streamflow information, typically have the least support from cooperators who currently support much of the network. The application across the nation of a single set of rules for locating gages may not adequately ensure streamflow information coverage in all situations. For example, the committee's analysis shows that NSIP sites are sparsely located in Nevada, to some extent because many of Nevada's streams are ephemeral. It may occur that in the future, streamflow information from ephemeral streams has greater value than is presently perceived, as more people move west to states such as Nevada.

The principle of adaptive management should be incorporated explicitly into the NSIP program to periodically reevaluate the network to ensure that it meets anticipated future needs for streamflow information.

NSIP SUPPORT FOR RIVER SCIENCE

An understanding of the integrated hydrologic, geomorphic, and biological processes in rivers—here termed "river science"—is a prerequisite for effectively managing rivers for navigation, water supply, power generation, or ecological functions. As an example, the closure of Glen Canyon dam on the Colorado River in 1963 changed the magnitude, timing, and temperature of streamflow and reduced sediment inputs into the Grand Canyon segment of the Colorado River. This has impacted the number and sizes of sandbars that are used by river runners and form the habitat for native fish. The 1996 controlled flood released from Glen Canyon dam was an experimental effort to rebuild sandbars and evaluate the potential for controlled flooding as a management tool. The effectiveness of such management has to rest on the scientific understanding gleaned from river science. With the recent addition of the biological resources discipline to the water resources, geologic, and geographic disciplines, the NSIP can be an important contributor to river science at the USGS.

River science is intimately concerned with flow regime, sediment transport, and channel morphology and integrates information on streamflow, water quality, sediment load, and biota from headwaters to mouth. This requires the characterization of river systems continuously in space, not just at gaging stations, and would benefit greatly from a more comprehensive NSIP data delivery system, focusing on streambed, sediment, and velocity distributions, as well as the discharge itself. Data of relevance to river science should also be rescued from historical files and made available on the Internet; these include crest stage data, slope-area data from flood studies, gaging station channel geometry, and bed sediment characteristics.

The USGS should identify watersheds for which good hydrologic information is available and land-use changes are documented. These sites should be prime sites where hydrograph information is retrieved and stored to better understand how changes in land use affect hydrologic characteristics. These issues are also being examined at experimental watersheds operated by other federal agencies, such at the U.S. Forest Service and the Agricultural Research Service. Close coordination with the efforts of these agencies and the academic communities that work at these sites is desirable.

With the addition of channel morphology data, sentinel (and other) watersheds can provide not only hydrologic reference points for the nation but stream morphology reference points as well. The representativeness of sentinel watersheds for characterizing the hydrologic and geomorphic diversity of the nation in support of river science should be explicitly evaluated.

1
Introduction

The goal of the U.S. Geological Survey (USGS) streamgaging program is to provide streamflow information to educate and inform resource managers and the public on defining, using, and managing water resources. The USGS meets this goal with a network of gages and with staff scientists and collaborators to study streamflow and river processes. There are many beneficiaries of USGS information because streamflow affects human safety, recreation, water quality, habitat, industry, and agriculture. A short list of applications noted by users in Illinois (Knapp and Markus, 2003) included assessing cultural resources, biological and conservation assessment and instream flow needs, current operations of water resources, floodplain mapping, hydrologic and hydraulic design and modeling, legal obligations, long-term flow statistics, recreation, regional hydrologic analysis, river forecasting and flood warning, water quality analysis, water resources operations planning, and education.

However, the streamgaging program is now challenged to adapt to changing economic conditions. Funds are tighter, even as the U.S. population grows, stressing water supplies, affecting ecosystem health, and moving into marginal flood- or drought-prone areas (Figure 1-1).

Today, a mix of funding from federal, state, and local agencies supports the USGS streamgaging program. The vast majority of this funding (93 percent) comes from partnerships with state and local agencies through the Cooperative Water Program (*http://water.usgs.gov/coop*) and with federal agencies such as the U.S. Army Corps of Engineers and the Bureau of Reclamation (Figure 1-2). Partners (or "cooperators") support streamgaging to obtain streamflow information that meets their needs; streamflow data from these streamgages also produce information that helps meet the

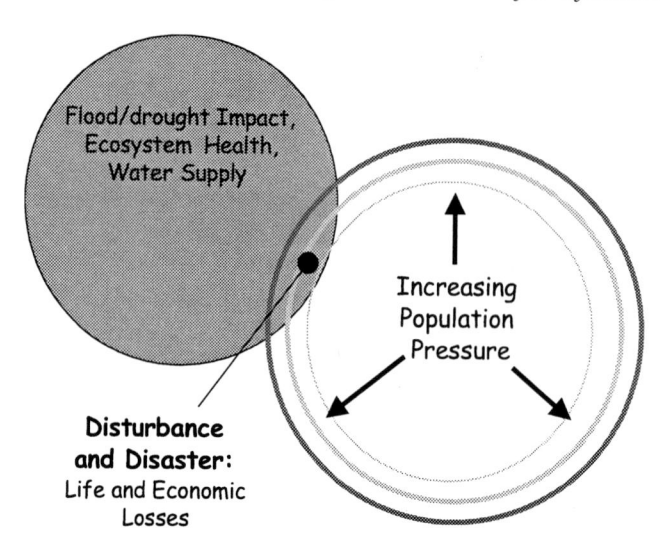

FIGURE 1-1 Life, economic, and habitat losses from increases in population near a river and within its watershed. The concentric circles are designed to show how increasing population begins to put pressure on other resources that were reasonably compatible with a smaller population. SOURCE: Adapted from USGS (*http://marine.usgs.gov/fact-sheets/nat_disasters/Circles.gif*).

broader needs of the nation as a whole. This means that the siting of streamgages is driven more by the needs of partners than by an overarching plan for meeting the nation's need for streamflow information.

The USGS reported that the ability to meet federal streamflow information needs had been degraded because of (1) a decrease in the number of streamgages, (2) a disproportionate loss of streamgages with a long period of record, and (3) the declining ability of the USGS to continue operating high-priority streamgages when partners discontinue funding (USGS, 1998). Congress had also expressed its concern about "the steady decline in the number of streamgaging stations in the past decade, while the need for streamflow data for flood forecasting and long-term water management uses continues to grow" (U.S. House Appropriations Subcommittee on Interior and Related Agencies, 1998).

Introduction 15

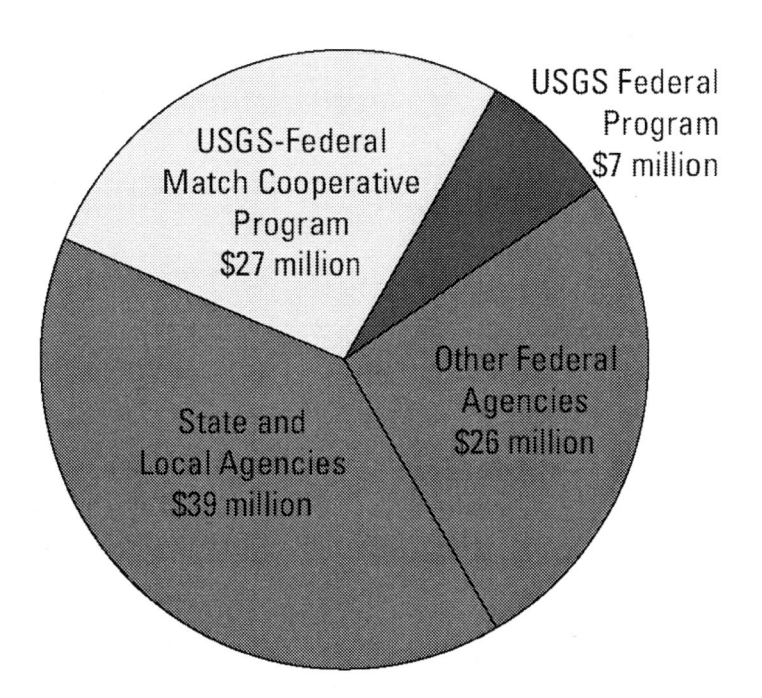

FIGURE 1-2 Fiscal year 2000 funding sources for the USGS streamgaging program ($99 million). SOURCE: USGS. (*http://water.usgs.gov/nsip/pubs/F-S048-01.pdf*).

THE NATIONAL STREAMFLOW INFORMATION PROGRAM

Recognizing the increasing needs for streamflow information, the USGS proposed the National Streamflow Information Program (NSIP) (USGS, 1999). The reference cited contains a general outline of the program, with specific numbers of gages recommended for different parts of the program. However, the present report may provide the most comprehensive description of the program that exists.

The NSIP plan has five components:

1. a nationwide system of federal interest streamgaging stations for measuring streamflow reliably and continuously in time;
2. a program for intensive data collection in response to major floods and droughts;

3. a program for periodic assessments and interpretation of streamflow data to better define their statistical characteristics and trends;
4. a system for real-time streamflow information delivery to customers that includes data processing, quality assurance, archiving, and access; and
5. a program of techniques development and research.

The streamgaging component of the NSIP proposal calls for a fundamental change in funding sources for the streamgaging program. The NSIP envisions a *federally funded base network* of streamgages designed to meet five minimum federal streamflow information goals for (1) interstate and international waters, (2) flow forecasts, (3) river basin outflows, (4) sentinel watersheds, and (5) water quality. A feature of the base network is the continuous, uninterrupted operation of its streamgages. Direct federal funding of these streamgages was proposed to remedy continuing losses of streamgages supporting these goals. The remainder of the USGS streamgaging network would consist of, as today, *cooperatively funded* streamgages. Cooperatively funded streamgages are designed to meet specific goals of federal, state, and local cooperators, and partnership with the USGS ensures that the streamgage data are quality controlled and available to all. Together, the base network and the cooperatively funded streamgages would meet many national needs for streamflow information (including the five federal goals).

The second component of the NSIP calls for intensive monitoring during times of major floods or droughts. Floods and droughts have serious social and economic impacts, including the loss of life and property, disruption of business activities, and interruption of water supplies. Intensive monitoring would include measuring streamflow where there are no permanent streamgages. Monitoring also would include gathering ancillary data on precipitation, river stage, and water quality. This component of the NSIP plan would support improved assessment of the risks, impacts, and mitigation of flood and drought hazards and provide new information for better scientific understanding of flood and drought processes and the effects of hydrologic extremes on river geomorphology and ecology. Much of the streamflow information generated by streamgaging results from careful analysis and synthesis of observations made at individual streamgages or a network of streamgages. The third component of the NSIP plan calls for periodic regional and national assessments of streamflow characteristics. Examples include regular updates of frequency estimates for low and high flows and regional synthesis to produce estimates of streamflow characteristics at ungaged sites. Assessments would also provide information on emerging scientific questions, such as the impact of climate variability on the magnitude and frequency of floods and droughts. The value of streamflow information derives from its use in decisionmaking and scientific inquiry. In recent years, the USGS began distributing streamflow information over the Internet, and there has been a dramatic increase in the use of real-time and historical observations by the public, water managers, and scientists, among others. Thus, the fourth component of the NSIP plan calls for enhanced delivery of its streamflow data and information products.

The fifth component of the NSIP plan calls for methods development and research for streamgaging. A significant portion of the annual cost of streamgaging is making direct measurements of discharge at gage sites to maintain the rating curve used to convert continuous measurements of river stage into streamflow estimates. Recent advances in technology have the potential to reduce the costs and increase the safety of making discharge measurements. These include acoustic Doppler technology to quickly make discharge measurement on large rivers and approaches that do not require sensor contact with the flow (for safety) and could potentially be made remotely (to reduce the need for site visits).

STATEMENT OF TASK

The National Research Council was asked to review the National Streamflow Information Program with respect to the following:

1. The minimum national streamflow information needs that should be met by the network, including those related to interstate and international waters, flood forecasts, river basin outflows, sentinel watersheds, and water quality.

2. The components of the NSIP plan that are reasonable, appropriate, and sufficient, including an enhanced nationwide streamgaging network with a larger share of national funding, intensive data collection during major floods and droughts, periodic regional and national assessments of streamflow characteristics, enhanced streamflow information delivery to customers, and methods development and research.

3. The ways a National Streamflow Information Program should support the data and information needs of various fields of river science, in addition to meeting its operational objectives.

ORGANIZATION AND CONTENT OF THIS REPORT

This report examines the goals of the NSIP to ensure that they are being reasonably and efficiently met. It evaluates streamgage network design, node (gaging station) design, and information delivery to consumers. It further addresses the tools to optimize the network design to maximize its efficiency and national coverage of streamflow and the technologies to improve gaging station efficiency and utility. To this end, a broad view is used of what might constitute a gaging station. The report examines interagency collaborations to effectively add nodes to the network. It looks at the merits of considering the streamflow program as primarily an *information* program, (i.e., data acquisition and analysis and information delivery), rather than as primarily a data-gathering program. Finally, it examines how streamflow information is used by consumers, to ensure that the needs of the public and water managers are both being met. Given that the NSIP has many beneficiaries, the study also addresses who should support it. Specifically, is there a rationale for federal support of a program that traditionally has been supported in large part by cooperators and beneficiaries?

Chapter 2 reviews the history of streamgaging at the USGS and examines the rationale for federal involvement in streamflow information by comparison with practice in other countries. Chapter 3 examines each of the five criteria used to select NSIP base network gage sites and studies the distribution of gage locations across the nation resulting from these criteria. Chapter 4 looks at the question of where to site streamgages and how long such sites should be maintained. Chapter 5 focuses on the other data collection and information components of the National Streamflow Information Program. Chapter 6 introduces the subject of river science and places the subject of streamflow information onto a background of the geomorphology and biology of stream and river systems. Finally, Chapter 7 presents the committee's conclusions.

2
Gaging the Nation's Streams

The U.S. Geological Survey (USGS) has a long tradition of studying the nation's streams. The first USGS gaging station was established on the Rio Grande in 1889 (Wahl et al., 1995). However, since the USGS's inception, its mission and programs have sometimes come under scrutiny by Congress or by the USGS itself, and as a consequence the mission and programs have adapted to changing needs and mandates. The National Streamflow Information Program (NSIP), as it is presently known, is being examined at the request of the USGS with a view to ensuring that it meets the nation's needs.

In this chapter, the committee traces the history of river studies and streamgaging at the USGS, summarizes what a USGS gaging site generally looks like, briefly consider the role of other U.S. agencies in supporting streamgaging, looks at streamflow network design in other countries, and examines the value of a national streamflow information program.

A HISTORY OF THE STUDY OF RIVERS AT THE USGS

A brief history of river studies at the USGS, in its various manifestations, provides background for review of the program. The following discussion was gleaned from a more general history of the USGS's first century of operation (Rabbitt, 1989). The picture that emerges is that of a program that traditionally has provided information to a host of users, funded as much by users as by federal government appropriations. Information includes hazard (flood and drought) estimation and warning and water supply information for irrigation (food supply), power generation, flood control, defense, and resource protection.

The USGS mission when it was formed in 1879 was "classification of the public lands." The federal government owned more than 1.2 billion acres, most of it west of the Mississippi River, and less than 20 percent of this land was then surveyed for mineral wealth or agricultural potential. John Wesley Powell in 1878 showed that most of this land was arid, and only a fraction of that could be irrigated. Water was clearly the limiting resource for development of the arid region, so Powell recommended organizing the arid lands into irrigation districts.

Irrigation and flood relief were tied together in an investigation by the USGS into using flood-generating water surpluses from the Rocky Mountains to irrigate dry areas west of the Rockies. A drought in 1886 seized the nation's attention, and in 1888 Congress authorized a survey of the western lands for irrigation potential. Sites were to be selected for reservoirs for storing water and at the same time alleviating downstream floods. This congressional authorization gave Powell, then USGS director, a long-awaited opportunity to map watersheds and measure streamflow (Figure 2-1).

Powell wasted no time in starting, even though he had to train hydrologists. Land purchases were put on hold until Powell's irrigation survey was complete, in order to prevent land speculation. (Many parcels of dubious value might be bought up by a company that would reap large profits once a water supply was demonstrated.) Western developers, understandably, were unhappy. Six new states that were given "dowry" lands could not settle them, giving them no tax base. In 1890, in response to pressure from these states and the developers, Congress repealed the withdrawal of lands and discontinued the irrigation survey. The USGS fell out of favor with Congress, and the next few years saw cuts in appropriations except for activities of immediate practical use, such as mineral resources surveys. The Senate appointed a committee to investigate the "efficiency and utility" of the USGS, an action directed at Powell.

The USGS survived this scrutiny, and Powell's vision survived in the sense that geology now included the study of water. A small appropriation in 1894 was earmarked for "gauging the streams and determining the water supply of the United States." Groundwater and water-use investigations became part of the USGS, and appropriations were increased regularly. The federal need for water information was fully recognized in 1896, when a Public Lands Commission was recommended, to include the director of the USGS. This commission was to be responsible for determining, among other things, the water supply of the public lands.

Theodore Roosevelt outlined a water policy in his first State of the Union message in 1901. The Newlands Act in June 1902 promoted reclama-

FIGURE 2-1 Streamgaging by the USGS in 1890. SOURCE: Rabbitt (1989).

tion of the arid lands, and the Reclamation Service, then an adjunct of the USGS, was established. The USGS Hydrographic Division separated from the Geologic Branch and became the Hydrographic Branch. Appropriations increased for water resources investigations over the years, in response both to irrigation needs and to several major floods (Figure 2-2). Streamflow measurement and analysis came into its own, linked to the development of waterpower, irrigation, and flood hazard estimation.

Waterpower interests increased after World War I, when USGS engineers conducted a national survey for waterpower sites. In 1920, the Federal Water Power Act established the Federal Power Commission, which could license the development of waterpower on federal lands. The USGS was given the task of measuring streamflow and examining proposed waterpower projects (Figure 2-3).

Cooperative funding drove the majority of investigations. Cooperators included the U.S. Army Corps of Engineers, who needed streamgaging for flood control projects, and the Department of State, which had international water issues to resolve. States also became important partners during

FIGURE 2-2 Devastating floods such as this one in New Jersey, in 1902, highlighted the need for streamgaging for warning, stream studies, and hazard estimation. SOURCE: Rabbitt (1989).

FIGURE 2-3 A USGS geologist surveys a western river for power generation potential in 1920. SOURCE: Rabbitt (1989).

this period. In 1905, Congress appropriated funds specifically for cooperative studies, and in 1928, Congress gave formal recognition to the federal-state partnership that became the Federal-State Cooperative Water Program (now known simply as the Cooperative Water Program). While Congress increased the water resources funding at that time, it stipulated that the maximum federal contribution to such projects would be 50 percent. As discussed later in this report, this limiting stipulation has had a major impact on the design of a federal streamflow information program.

The Hoover presidency (1929-1933) was important for the USGS because President Hoover believed in both conservation of resources and basic research to understand them. For example, the destruction of ground cover by overgrazing had worrisome implications for water supply. In response to such concerns, the Water Resources Branch expanded. The depression heralded a sober era for the USGS in which basic research was conducted in the shadow of practical hydrology. Overall, Franklin Roosevelt's programs actually led to growth in the USGS. The Tennessee Valley Authority and the Public Works Administration both required intensive streamgaging, but the grants also supported research. The USGS made great strides in quantitative hydrology, researching rainfall-runoff relations and analyzing flood frequencies. Streamgaging instrumentation also improved. With these new program funds, federal appropriations now accounted for only one-quarter of total USGS support.

After World War II broke out, the USGS Water Resources Branch had responsibility for providing information on water for military and industrial installations. The USGS wrote more than 15,000 reports for the war effort. After the war, the USGS focus shifted back to irrigation, flood control, and highway drainage. The agency also took on the task of determining water needs for industry, starting with the steel industry. These tasks resulted in a very active USGS by 1954 (its seventh-fifth year), at which time 6,400 gages were active. New research activities in the next decade set the tone for the rest of the century. The USGS researched stream sediment transport, including measurement methods, bedload, controls on channel aggradation, and effects of sediment on flow. It conducted basic process research on river hydraulics in the field and in flumes, from the large scale (stream networks) to the reach scale, investigating relationships among discharge, channel geometry, drainage basin size, and water velocity. It initiated studies to answer management questions, such as effects of reservoirs on flow and impacts of wetlands and groundwater pumping on streamflow. USGS scientists also studied flood hazards; water supply issues, including water resource assessments; snowpack and snowmelt, annual runoff estimates for major basins; and effects of rainfall and drought on flow. By 1962, the Water Resources Division of the USGS was involved in "fundamental and applied research in water hydraulics, limnology, hydrology of ground water and surface water, geochemistry of water, stream-channel development and morphology, sediment production and transport, evapotranspiration and evaporation suppression, physical and chemical interrelations of precipitation and water above and below the land surface, and the effects of man-made environmental changes on water and water supplies" (Swenson, 1962).

The 1960s onward could be characterized as an era of USGS participation in public issues. As the nation began to confront its industrial and radioactive wastes and their human health hazards, the USGS took a larger role in these areas as well as natural hazards. Geochemists shifted from mainly mineral prospecting to exploring the distribution of potentially hazardous natural substances. The 1964 federal budget gave the USGS the task of creating a national network for collecting water data to address accelerating demands on resources and movement of Americans into water-poor or flood-prone areas. The goal was a 50 percent increase in collection of basic water data by 1973. The network would be supported by the development of digital recording equipment, computerized data processing, and central data distribution through the new Office of Water Data Coordination. The value of basic research was also emphasized, and the plan called for a water-resources program that was 25 percent research. Scientists were needed to do this work, so the USGS helped develop hydrology curricula at major universities. Although the Vietnam War pulled attention and resources away from many domestic programs, environmental problems stayed in the public eye. The 1960s and 1970s saw passage of the Water Quality Act, the Solid Waste Disposal Act, and the National Environmental Protection Act.

By 1971, the USGS collected streamflow data at more than 11,000 gaging stations and measured water quality at 4,000 stations. Multidisciplinary studies had increased in number, and information became increasingly accessible. Hazard prediction (including flood prediction) was given high priority. A technological breakthrough came in 1972 with the availability of what is now called Landsat satellite data. The USGS built a data center in South Dakota to distribute satellite and other remotely sensed data and immediately began exploring how the new information might address hydrologic issues. In 1975, the Land Information and Analysis Office consolidated several multidisciplinary land resource and environmental programs. One of its main objectives was to interpret and display land resource information for a wide audience.

In 1977, the National Water-Use Information Program was created, and its five-year reports continue to be the most widely used USGS products. In 1984, the program also started publishing the National Water Summary, which annually described hydrologic conditions and events (such as floods) for each state.

The 1980s were a time of downsizing and increased private access to federal lands for mineral and energy development, in order to increase domestic energy and mineral production. As a result, the USGS reverted to its initial task of classifying public lands, and some of its other duties were placed in other agencies. The primary task of the Water Resources Division was to provide hydrologic information for the best use and management of water resources. Mapping advances benefited the Water Resources Division; by 1988, the Mapping Division completed the 1:100,000-scale digital database including hydrology of the United States.

USGS publications from the last few decades reflect emerging technologies and changing societal values, linking streamflow to water quality, land use, and watershed management. Desired flow characteristics reflect changing values and are increasingly related not just to power supply, flood protection, and human water supply, but also to biological functions of rivers, including riparian habitat. The USGS has also taken advantage of technological breakthroughs in computational capacity, satellite communications, geographic information system (GIS) technology, and remote sensing. Computer flow models are used to estimate sediment transport, estimate streamflow highs and lows from precipitation, extend flow records, reconstruct natural flows, forecast future water demand, and predict effects of climate change on streamflow. The USGS has paid considerable attention to the statistics of streamflow and has developed field methods and mathematical tools to minimize the uncertainty of its numbers. It has also sought to make its information rapidly and readily accessible to the public, especially through the Internet.

The present-day NSIP developed in response to critical national needs—irrigation water supply (with national interest heightened by severe drought), flood warning and flood estimation, public water supply, waterpower generation, water conservation, national defense, and industrial water supply. Now the streamflow program serves the additional needs of protecting water quality and aquatic and riparian habitat, watershed management, and providing information for river recreation. A tension has always existed between applied hydrology to provide specific kinds of information for a specific purpose at a given location, and basic hydrologic science to understand streamflow. Project-based funds have been augmented to a greater or lesser degree by federal appropriations that in some cases could serve basic research needs. Whenever possible, the USGS has strived to maintain hydrologic research in the interests of the long-term water supply and hazard prevention.

The next two sections discuss the streamgaging network and its "nodes," the individual gaging stations that define the network. It should be noted that NSIP is not the only network within this larger set of gages. Other important networks include gages used in three streamflow and water quality networks: the National Water Quality Assessment (NAWQA) program, National Stream Quality Accounting Network (NASQAN), and Hydrologic Benchmark Network. The component gages of these and other networks overlap with those of NSIP and each other. Thus, the NSIP network includes gages funded by these and other programs, including those supported by matching funds provided by other federal, state, and local agencies. Data from all USGS gaging networks are gathered into the National Water Information System (NWIS) database of the USGS, accessible via NWISWeb.

WHAT IS A GAGING SITE?

The USGS's stream science program rests on the data collected with the streamgage network of about 7000 gages. A streamgage's main purpose is to measure a river's *discharge*. Recorded as a volume of water per unit time (usually in cubic feet per second), the discharge is crucial information about water available for drinking, irrigation, industry, energy, engineering, recreation or wildlife, or on the other hand, the downstream flood risk. River discharge is labor-intensive to measure, so gaging stations instead record a river's water level, or *stage*. Changes in stage originally were recorded by using a float attached to a rotating drum and, more recently, have been recorded by using pressure transducers that convert water pressure to an electronic signal. A sturdy housing protects most USGS gages; even during severe floods the gages must continue to function and transmit information or they lose their value for flood warning.

Stage is then converted to discharge with a *rating curve*. Building the rating curve is part of the cost of streamgaging, because discharge measurements must cover the whole range of stages that a river might reach. USGS personnel must visit the gaging station numerous times at various discharges and measure both stage and discharge directly. Discharge is typically measured with a current meter (Figure 2-4). The river width is divided into intervals, and for each interval the water depth and a representative water velocity (usually the velocity recorded at 60 percent of the total depth) are measured. Multiplying the area of each interval (square feet) by the velocity (feet per second) provides a discharge for each interval; the sum of these is the total discharge for the river.

FIGURE 2-4 Measuring discharge by means of a bridge crane. The current meter, or "fish", is lowered into the river to measure current velocity. The crane is wheeled along the bridge to obtain measurements at multiple sections. SOURCE: USGS (*http://water.usgs.gov/wid/FS_209-95/mason.figure.id.1.gif*).

Such direct measurements of discharge are consistent and robust. They have not changed fundamentally in a century. They have the disadvantage that for practical reasons, flows cannot be measured at every possible point of interest within the river system. The theory to extrapolate flows from measured points to other points of interest is poorly developed. An opportunity exists to put flow estimation on a more theoretical footing by constructing numerical models of streamflow hydraulics at gaging station sites.

The rating curve may shift with time in channels that are eroding or aggrading. Shifting rating curves introduce error into discharge measurements. To minimize such errors, the USGS attempts to locate gages at relatively stable *control sections*, such as near bridges. In general, however, channel sedimentation or erosion can be expected, so the USGS must make frequent measurements, especially at high discharges, to keep the rating curve up-to-date or it loses its value.

As might be expected, very high stages and discharges are rare but are of great interest for flood warning. The USGS strives to amass data on high discharges whenever and wherever they occur, in order to extend rating curves into the high-flow range. These high-flow conditions are hazardous, so techniques and tools continue to be developed to keep USGS personnel out of harm's way.

Once the rating curve has been constructed, raw continuous measurements of stage are transmitted to the USGS, where they are aggregated, converted to periodic discharge, and delivered in real time to users via the Internet. Not all data are disseminated: many are archived by the USGS, either digitally or otherwise, including notes by field hydrologists, rating curves, and so-called unit values of discharge.

Gaging and data retrieval innovations have led to variability among the 7000 USGS gage stations. The simplest gage station may be a temporary or one-time measuring point, in some cases simply a tube filled with cork crumbs to record the highest stage by leaving a bathtub ring of cork in the tube. The "crest stage" so measured is increasingly seen as a biologically critical streamflow parameter (e.g., Bovee and Scott, 2002; Scott et al., 1997). Other gage stations consist of webcams and simply show hourly photos of flashflood-prone rivers such as the Santa Cruz River in Arizona (Figures 2-5 and 2-6). At the other end of the spectrum is the fully automated multi parameter gage station that transmits data in near real time from a remote location via satellite (Figure 2-7). The great majority of USGS gages are now equipped with these systems. Data are transmitted by two geostationary operational environmental satellites (GOES) operated by the National Oceanic and Atmospheric Administration (NOAA). Data are retransmitted by domestic satellite to the USGS and other users.

The hazards associated with streamgaging and the need for intensive data collection during rare high-discharge events have led the USGS and others to develop "non-contact" technologies, such as pulsed Doppler radar to measure surface velocity, and ground-penetrating radar to measure channel cross section (Costa et al., 2000; Haeni et al., 2000; Melcher et al., 1999; and Spicer et al., 1997; also see Chapter 5 of this report). These technologies can be deployed at a particular station or on a mobile unit for measuring conditions at many stations during a high-flow event. Thus far, they have not been widely used (Table 2-1).

The preceding discussion raises the question of whether the existing gages are technologically optimal. Are national needs being met at critical sites? Can innovation reduce long-term labor costs? Some of the issues that face the USGS in effectively gathering streamflow information are listed below, and several are discussed in more depth in later sections.

- _Personnel_: need for frequent site visits to build and update rating curves, with an even greater need during large regional floods

FIGURE 2-5 Gaging station on the flashflood-prone Santa Cruz River in Arizona includes a webcam to transmit hourly photos to warn of floods in the otherwise dry channel. SOURCE: USGS (*http://az.water.usgs.gov/webcam/9482500_cam/cam_09482500.html*).

FIGURE 2-6 Arizona's Santa Cruz River, normally dry, in flash flood, 1983. SOURCE: USGS (*http://az.water.usgs.gov/webcam/9482500_cam/cam_09482500.html*).

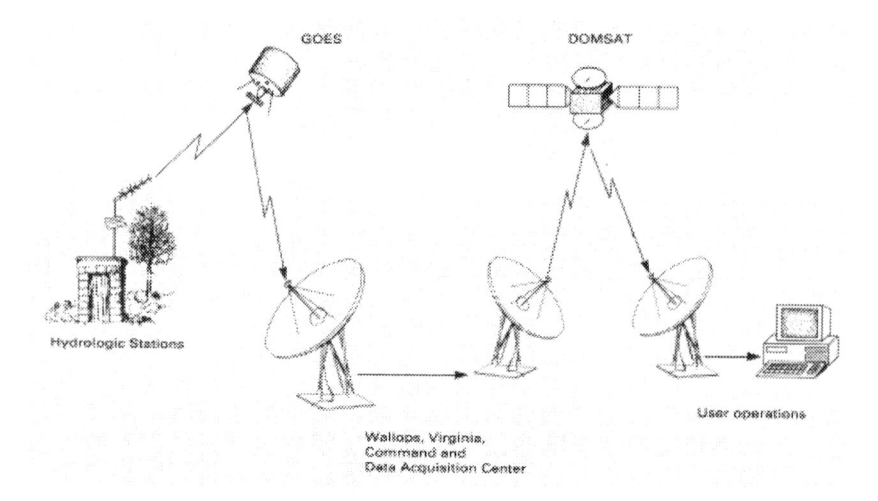

FIGURE 2-7 Most USGS gage stations transmit data on river stage in real time, using two satellite links. SOURCE: USGS (http://md.water.usgs.gov/publications/presentations/md-de-dc_rt98/sld025.htm).

- *Safety*: need for technology to measure discharge quickly and remotely
- *Communications*: need for information to reach the affected public quickly, despite possible interruptions in communication lines
- *Durability*: need for gages and transmission devices to continue to function even in severe conditions
- *Water supply security*: need for information on *low-flow* conditions to provide decision makers and water managers with information to manage needs for drinking water, power generation, recreation, defense, industry, and instream habitat
- *Distribution and coverage*: need for knowledge of conditions at any time, whether measured directly or interpolated
- *Continuity*: need for long records in order to understand extreme events and assess stationarity of streamflow
- *Cost optimization*: need to optimize the balance between spatial coverage and long records, given that resources are limited

TABLE 2-1 Summary of Hydrologic Stations [a]

	Feature	Number Active	Percentage of Total	Number Inactive	Costs Technology	Labor
Data collected	Continuous stage	7,273		12,151		
	Crest stage only	[b]		[b]		
	Discrete (event) data	[b]		[b]		High
Data retrieval	By site visit	1,260	17.3		Low	High
	By satellite	6,013	82.7		High	Low
	By camera	0	0		High	Low
Remote sensing	Ground-penetrating radar	0	0	0	High	
	Doppler	200-300			High	

[a] The numbers of gages recording stage maxima or discrete events are not tracked because of their inherently changing nature.

[b] Data on numbers of these nonstandard measurements are not readily available. SOURCE: J. Michael Norris, USGS, written communication, March 2003.

THE NSIP GAGING NETWORK

A discussion of streamgaging must include not just *what* is measured and *how* it is measured, but *where* it is measured. The benefits provided by gages exist only where the network covers a particular area. The loss of a gage may represent an information loss to the network, but perhaps more critically it represents a loss of coverage for certain communities or for certain gaging needs.

The USGS, faced with constraints, has designed the NSIP to provide full coverage for certain needs (e.g., interstate compacts, compliance with international water treaties, estimating major river basin outflows). The prioritization that the USGS appears to have used, even if not stated explicitly, has been not by gage but by federal gaging needs. The question has not been, Do we need this additional gage? but Do we need this kind of coverage? This question is examined more closely in Chapter 4.

If one views the gaging network as a coverage problem, locating a streamgage at a site is just one way of achieving coverage. Periodic site visits, temporary gages, statistical estimation, GIS models, or other new technologies might also achieve coverage. What is needed is coverage, not gages per se. A more detailed discussion of principles and trade-offs of streamflow network design is contained in Chapter 4.

Another aspect of the gaging network is that it is reassessed periodically. Gaging is therefore an example of adaptive management, in which the fundamental goal is to obtain coverage either directly or indirectly for the priority gaging needs.

ROLE OF OTHER AGENCIES IN SUPPORTING STREAMGAGING

Many city, county, state, and federal agencies collect streamflow data. The primary differences between USGS networks and those of the agencies are the purposes for which data are collected. Other agencies generally collect only those data needed for a specific mission or task. For example, data collected to fulfill wastewater permitting requirements often do not include the full range of flows. These data, while vital for their own goals, are generally of limited value in addressing issues of national and regional scope (Hren et al., 1987). As a result, these data are usually not placed in accessible archives and made readily available. One possible solution would be for the USGS National Water Information System to contain pointers to sources of streamflow information other than those contained in USGS archives.

Some data collected by other agencies, however, have value beyond the specific purpose for which they were collected. Data from some stations operated by state and federal agencies are quality assured by the USGS, published in the annual state Water Data Reports series compiled by the USGS, and entered in the USGS database. In 1990, data from about 400 stations were provided to the USGS by other agencies (J. Michael Norris, USGS, written communication, 2002). In fact, the many interests served in federal programs (e.g., the Environmental Protection Agency's [EPA's] Total Maximum Daily Load [TMDL] program and the Federal Emergency Management Agency's [FEMA's] Flood Insurance Program) by the USGS streamflow information are a strong argument for federal support of the NSIP.

STREAMFLOW NETWORK DESIGN IN OTHER COUNTRIES

It is useful, in assessing the way in which the United States gathers and disseminates streamflow information, to look at how other countries manage the collection and dissemination of this information. Examples of some of these arrangements are summarized below.

Australia

In Australia, the responsibilities for water resource assessment and management are vested in the states under the Constitution of the Commonwealth, and state or territory governments currently fully fund these networks (Ross James, Commonwealth Bureau of Meteorology, personal communication, 2002). Only the climate and weather networks operated by the Australian Bureau of Meteorology are maintained with Commonwealth funding because meteorology is a Commonwealth responsibility. The bureau also provides a national flood warning service under collaborative arrangements with state or territory and local governments. As a result of these arrangements, the bureau does operate some stream monitoring stations. However, the state or territory and local governments operate the majority of stream stations used by the bureau's flood warning service.

Up until the mid-1980s, some Commonwealth funding was provided to the states for streamgaging networks. An attempt to identify specific stations that made up a national monitoring network for which funding would be provided resulted in Commonwealth funding support being redirected toward "project-based" initiatives rather than a national monitoring system.

Currently, Australia is in the process of completing a National Land and Water Resources Audit (*http://www.nlwra.gov.au/*), which is funded by the Commonwealth government with considerable matching support from the states and territories. The need for improved monitoring, ongoing monitoring, consistent data management standards, and improved access to data and information products has featured prominently in audit reports. These issues, and the role the Commonwealth government will play in addressing them, still have to be addressed as part of the plans for ongoing audit activities.

National streamgaging information is available on-line at the Bureau of Meteorology site as a catalog of the water quality monitoring stations operated by the state and territory water agencies. However, only descriptions of the data are provided. The observations on streamflow must be obtained from the agency operating the station.

Canada

Canada's Hydrometric Program is carried out under formal agreements (signed in 1975) between Environment Canada and each of the provinces and Indian and Northern Affairs Canada, representing the territories under the Canada Water Act. The agreements provide for the collection of surface water quantity and sediment data on a national basis, and the costs of the program are shared according to specific interests and needs. Over the years, a number of interpretations of the agreement articles have occurred. Currently, the program operates 2,290 water-level and streamflow stations. An additional 413 stations are operated outside of the program (Table 2-2).

According to national guidelines for designating water quantity survey stations, federal stations (i.e., those funded 100 percent by the government of Canada) support programs of primary interest to Canada which include the following:

1. *Federal Departmental Programs.* These are stations required under statutory obligations that have developed in response to federal legislation and priorities and as a result of programs of various federal government departments or agencies to provide quantity information on inland waters. They include stations operated in support of specific federal works, benchmark basins, studies or investigations, and research projects and to meet navigational requirements and management responsibilities. A station may be so designated where Canada has formally accepted responsibility for continued operation of the station under an implementation agreement.

TABLE 2-2 Canada's Streamgaging Network

Category (funding)	Number of Stations	Percentage of Active
Federal	671	25
Federal-provincial or federal-territorial	863	32
Provincial or territorial	756	28
Fully cost-recovered from other parties	94	3
Contributed by other organizations	319	12
Total active stations	2,703	100
Total inactive stations	5,300	—

SOURCE: Environment Canada.

2. *Interprovincial Waters.* These are stations required for monitoring waters flowing across or forming part of provincial or territorial boundaries where federal responsibility has been established by an agreement or justified by an interjurisdictional concern.

3. *International Waters.* These are stations associated with federal responsibilities arising from international agreements, treaties, orders, or studies, including the following:

• Stations specifically named under the Boundary Waters Treaty and those approved officially as "international gauging stations"
• Stations specifically stipulated under International Joint Commission Orders, or required to support such orders, to provide for control of waters crossing or forming part of the international boundary and for International Joint Commission related study, surveillance, flow regulation, or apportionment purposes; such stations may also be required for similar studies carried out under unilateral or bilateral mechanisms and undertaken in anticipation of the need for formal orders
• Stations related to international treaties and agreements that involve waters crossing or forming part of the international boundary and specifically stipulate the reaches of streams required to be monitored or special arrangements that have to be made to meet water quantity survey needs
• Stations on streams flowing across or forming part of the international boundary for which Canada has determined that monitoring is required for water management purposes

4. *National Water Quantity Inventory.* These are stations that provide information for a national inventory of surface waters. They consist of those stations required to determine water quantity trends in the major drainage basins in Canada that serve to provide an assessment of the total surface water resources and to measure significant discharge to the oceans.

In many respects, the Canadian program resembles the U.S. program.

United Kingdom

The United Kingdom maintains a network of more than 1,300 gaging stations. Responsibility for these stations rests primarily with the Environment Agency in England and Wales, the Scottish Environment Protection Agency, and in Northern Ireland, the River Agency. The data are archived by the Centre for Ecology and Hydrology with funding from the Natural Environment Research Council.

Brazil

The federal government of Brazil provides 100 percent federal funding for 5,000 stream gages as a part of the water quantity and quality monitoring program. All hydrologic data obtained through this program are made available free of charge to all interested parties and individuals. The collection of the related meteorological data is also fully funded by the federal government, and administered by the Meteorology Institute of Brazil. However, the meteorological data are not available free of charge because the institute requires additional income to support its operations. The issue of charging for the meteorological data is subject to some debate within the Brazilian federal government.

Germany

In Germany, three institutions or organizations that are responsible for the streamgages (H. Gerhard, 2002; Hessian Agency for the Environment and Geology, personal communication, 2002; A. Sudau, Bundesanstalt für Gewässerkunde, Referat Geodäsie, personal communication, 2002):

(1) the federation represented by the Federal Waterways and Shipping Administration,
(2) the federal states (the Länder), and
(3) regional water associations and communities (used for dams and water works).

The legal basis is the Water Management Act (*Wasserhaushaltsgesetz*). There are 260 federal streamgages in Germany that are fully funded by the German federal government. The Federal Local Waterways and Shipping Offices operate these gages. The other gages are funded either by the federal states or by contributions to the associations. All data (such as high or low waters, mean daily or yearly discharges or water levels) are published in books related to the large rivers (e.g., the Rhine River Hydrologic Yearbook).

Currently, there is a federation committee that deals with the problem of optimization of gaging station networks in Germany. However, the main task of this committee is to optimize gaging networks in coastal areas, which include tidal rivers and estuaries. The committee developed a small brochure, but it is available only in German. The committee also reviewed the literature on network design and found that the majority of literature comes from the United States and was generated during the 1970s. The review of literature is also available in German (C. Blasi, LAWA Committee for Developing Criteria Catalogue of Gauging Stations in Coastal Areas, personal communication, 2002).

In summary, the streamflow information programs in other countries show that there is recognition worldwide of the vital importance of streamflow in serving public interests. Other countries have greater streamflow information coverage, in some cases because population densities have exerted greater pressure on resources than in the United States. Yet Canada, with a lower population density, has better coverage. The Australian case is particularly interesting because the Bureau of Meteorology provides a federal link to valuable streamflow data from states and territories.

European Environmental Agency

The design of a water resources monitoring network for the European Environmental Agency (EEA) (Nixon, 1999) identified seven different types of monitoring stations related to the type of information provided. These also correspond closely to the NSIP network design goals. In considering European Union (EU) water quality monitoring needs, the possible station types identified by the EEA:

- statutory stations to provide data for legal commitments, either regulatory, international transboundary waters, or obligations from EU directives;
- benchmark (or reference) stations to characterize catchments undisturbed by man;

- boundary stations to characterize fluxes at legal boundaries or across media;
- impact stations aimed at controlling human impacts associated with well-defined pollution sources;
- representative stations to provide summary information on larger areas with long records;
- operational stations for day-to-day management by local, regional, or national agencies; and
- research stations installed and operated during scientific projects.

Three general types of water quality monitoring stations were judged most relevant to the EEA monitoring network:

1. reference stations, supporting the analysis of natural or pristine water quality and trends across Europe;
2. flux stations; and
3. representative stations.

Additionally, two broad categories of stations were identified for inland water quantity monitoring:

1. statutory and operational monitoring to provide information for the business and operational needs of regulators, suppliers, and other users; and
2. surveillance monitoring to characterize and allow appraisal of the state of water resources and, with water quality and biodiversity information, the state of the EU water environment.

Surveillance monitoring stations include:

- reference stations that characterize undisturbed conditions;
- baseline stations that capture regional hydrology to characterize ungaged sites;
- representative Stations with long records to support regional and national assessments; and
- impact stations selected to characterize the effects of man's interference with the natural regime.

Motivated by a very different set of institutional drivers (such as EU directives), the station types identified for an EEA monitoring network are nonetheless quite similar to the goals proposed for the NSIP streamgage network. Although the EEA is not a primary collector of data, the information sought from the EEA monitoring network reflects EU member nations' need for unbiased scientific information to support assessment, management, and policy making—a need mirrored in the United States.

VALUE OF A NATIONAL STREAMFLOW INFORMATION PROGRAM

Four areas in which streamflow information clearly has value to society are (1) optimizing hydropower and water supply, (2) reducing impacts of flooding, (3) reducing impacts of droughts, and (4) reducing pollutant loads to waterbodies. Other areas where streamflow data can have high value include national defense, food and fiber production, recreation, and wildlife habitat and diversity including Endangered Species Act requirements. The relationship of streamflow information to aquatic habitat is examined in Chapter 6. The formal definition of information gain from gaging, and how it can be valued, is addressed in Chapter 4.

Optimizing Hydropower and Water Supply

An analysis in New South Wales, Australia, showed that the benefit of streamgaging in aggregate is about ten times the cost involved, but may be hundreds of times the cost for particular gages where water storage or flood mitigation is planned (Cloke and Cordery, 1993; Cordery and Cloke, 1992). In terms of power generation benefits on the Columbia River, long-lead streamflow forecasts allow alternative operation of reservoirs for hydropower production that result in an increase in $153 million per year in generation revenues (Hamlet et al., 2002). Streamflow data are critical for water management, allowing flow-based quantification of the dollar value of alternative uses of stream water (recreation versus municipal use versus power generation versus agriculture) (e.g., Bosch, 1991; Douglas and Taylor, 1998; Hansen and Hallam, 1991; Leones et al., 1997). Similarly, streamflow information enables the agricultural community to make economically sound decisions.

Reducing Impacts of Flooding

Flood disasters have a devastating impact on human lives and property. The National Flood Insurance Program operated by the FEMA has the mission of mitigating flood losses through insurance payments for flood damage. As shown in Figure 2-8, the number of flood insurance policies has increased steadily through the years; the number of damage losses paid out fluctuates significantly from year to year, averaging about 40,000 losses paid out per year in recent years; and the dollar value of these losses also varies significantly from year to year, averaging about $1 billion per year in recent years.

Generally speaking, streamflow data, including data uncertainty, are necessary for rational economic decision making for flood warning (Krzysztofowicz, 1999). The USGS has the federal responsibility in the United States for streamflow measurement, and the National Weather Service (NWS) has the responsibility for streamflow forecasting. Thus, the USGS is responsible for records of historical flows, and the NWS for forecasting future flows. These two responsibilities intersect in the present, where the National Weather Service uses real-time and historical streamflow information from the USGS in its flood forecasting operations. Although the number of gages in the national streamgage network has diminished slightly in recent years to less than 7,000, Figure 2-9 shows that the proportion of gages with satellite telemetry to transmit data in real time is increasing steadily, to currently more than 6,000 gages. Streamflow information in real time is critical to flood mitigation and forecasting efforts. It is very difficult to quantify the lives or property saved by a single gage used in a flood forecasting system. Without a doubt, gages are extremely valuable, but their value is encapsulated in the operation and accuracy of the entire forecast system, the forecast delivery mechanisms, and the flood forecast response.

Besides flood forecasting, streamflow information is also used in creating FEMA floodplain maps and, thus, in protecting property from flooding through flood ordinances. Most river reaches for which flood maps are constructed do not have streamgages on them, and flood peak estimates are defined by rainfall-runoff modeling. Streamflow information is used to calibrate the rainfall-runoff model at gaged sites in the flood study region and, thus to create confidence that the flood peak estimates defined for ungaged reaches are reasonable.

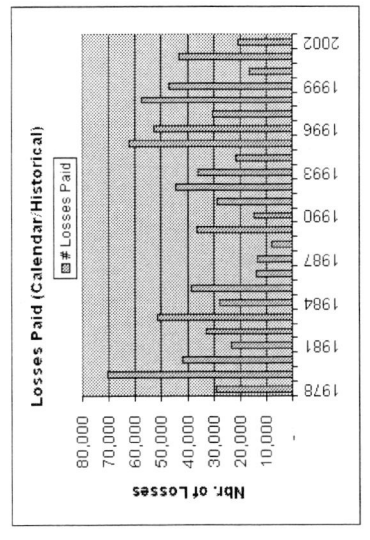

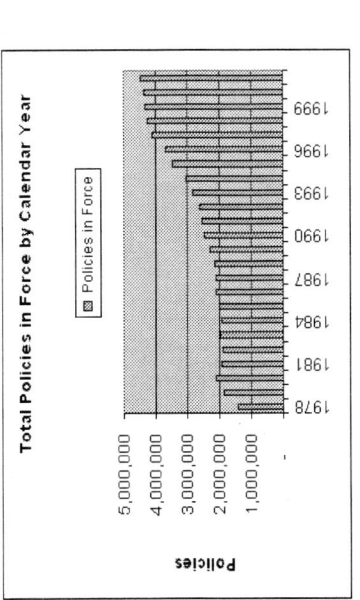

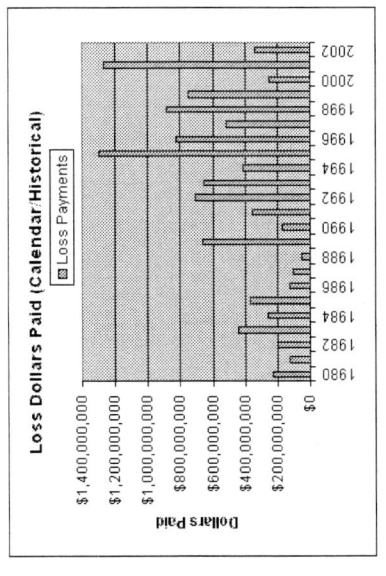

FIGURE 2-8 Trends through time in the National Flood Insurance Program. SOURCE: FEMA (2003; *http://www.fema.gov/nfip*).

USGS Streamgaging Stations

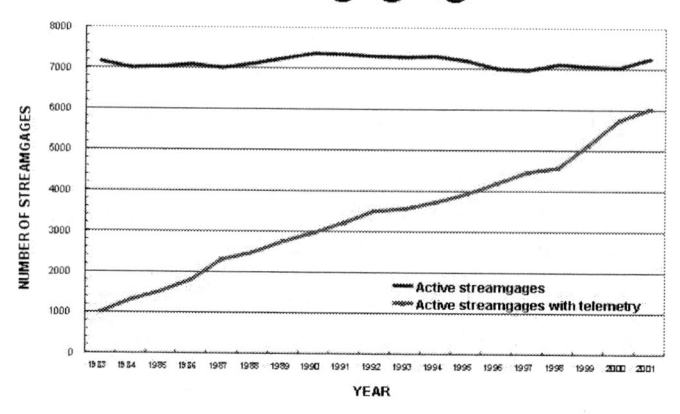

FIGURE 2-9 The total number of USGS gaging stations has changed only slightly since about 1990, but almost 90 percent of gages now have real-time data delivery, generally using satellite telemetry. SOURCE: J. Michael Norris, USGS, written communication, September 2003.

Reducing Impacts of Droughts

Periodic droughts dominate the water supply strategies in the arid western states. For many years the only offsetting action for droughts was thought to be the construction of increased dam capacity. Water supply management during recent droughts in the western United States has strengthened the realization that more precise streamflow forecasts and predictions can partially substitute for increased structural supplies. By making better use of existing storage capacity and allowing more precise regulation of minimum streamflows to meet environmental standards, better information can substitute for structures at a substantial saving. Essentially, the management of water supplies under drought conditions requires stochastic, dynamic decision making. That is, it can be demonstrated that given a supply safety standard defined as the probability of a certain level of shortfall, the greater the variance of future stream inflows to a dam, the larger the "safety stock" must be to ensure a given supply probability. The same logic applies to meeting environmental goals that are often defined in terms of minimum streamflow levels to protect endangered species. Better monitoring of the watershed streamflow enables more precise real-time prediction of the run-off as a first indicator of the severity of a drought. In addition, past monitoring information can lead to improved predictions of changes in streamflow needs during periodic droughts.

Recent droughts in the western United States have shown that both water supply and environmental water requirements can be managed more precisely with improved predictions and forecasts. Improved forecasts of water demand enable managers to enter into contracts for water transfers that are conditional on streamflow conditions. Such contracts enable water demand to be more flexible and to adjust to fluctuations in supply while maintaining supply reliability. However, these contingent transfer contracts depend on reliable forecasts of water demands under different streamflow conditions and on the ability to accurately monitor real-time streamflow conditions during droughts.

Reducing Pollutant Loads to Waterbodies

Water quality is also intimately linked to stream discharge and velocity, and discharge estimates are critical to accurate contaminant load estimates and pollutant reduction plans. Aside from the obvious fact that loadings are calculated as discharge times concentration, the sediment transport capacity of a river is highly dependent on velocity. In addition to sediment pollution itself, many inorganic and organic species (e.g., phosphate, heavy metals, pesticides, PCBs [polychlorinated biphenyls]) are attached to suspended clays, iron oxyhydroxides, and organic matter. As an example, USGS estimated the load of nitrogen to the Gulf of Mexico (Goolsby and Battaglin, 2000), an issue that bears on hypoxia and the loss of fisheries in the Gulf. Estimates of loads using nutrient inputs to the land (e.g., fertilizer use) were greatly improved by factoring in the stream discharge. This approach also suggested where nutrient management could most effectively be targeted (i.e., Illinois, Iowa, northern Indiana) to reduce loads to the Gulf.

Many recent environmental regulations have been promulgated as restrictions on the TMDL for a body of water or section of a stream. Total Maximum Daily Loads were established in the 1972 Clean Water Act. The TMDL is a measure of the assimilation or dilution capacity of the waterbody for a particular pollutant. Most causes of quality impairment fall into five categories: sediment and siltation, pathogens, metals, nutrients, and organic enrichment. From 1996 to 1999 there were only 300-500 TMDLs approved nationally, but approvals in recent years have ranged from 1,100 to 2,500.

TMDLs cannot be set accurately without reliable information on the characteristics of the flow in the waterbody. Clearly, the assimilative capacity of a waterbody is related to the average flow and its variability. Historical streamflow monitoring data are required to establish TMDL levels for different flow regimes and to determine when the streamflow is at the specified stages for different TMDL levels. Often, a single level or threshold is established due to a lack of detailed streamflow monitoring data. Prudence requires that single threshold TMDLs be set at levels that do not compromise the quality of the water at low-flow levels; however, these TMDL levels may have an unnecessarily high cost at other flow levels. Therefore, there is a direct inverse relationship between the precision of streamflow information and the efficiency and social cost of TMDL regulations.

RATIONALE FOR FEDERAL SUPPORT

The rationale for the National Streamflow Information Program rests on both the value of streamflow *information* and the *national* need for this information. Streamflow information, like most goods and services, can be provided through a variety of administrative and institutional mechanisms. Many public (e.g., flood control districts) and private (e.g., power generators) entities invest in streamflow information to satisfy their specific needs and applications. Private sector streamgaging is a common value-added service offered in association with environmental assessments and site evaluations. The common provision of streamflow information by the private sector naturally requires us to consider the national interest in streamflow information: Who benefits from streamflow information? Who should bear the costs?

Public Investment in National Streamflow Information

Streamflow information has many features of a product that is a "public good," serving the national or regional interest. Public goods are characterized by (1) the inability to exclude those who have not paid for the service, (e.g., radio broadcasts warning of floods) and (2) a zero marginal cost of servicing additional individuals. Because of these two characteristics, they are rarely provided by private enterprise. A survey of the main characteristics of and literature on public goods can be found in Kolm (1988). In his survey, Kolm stresses that the exclusion and marginal cost characteristics noted above are rarely absolute or "pure." In reality, the degree of exclusion and marginal cost extend from the pure public good, such as defense, to private goods. The defining factor is the cost of exclusion and provision. Information, in the form of streamflow data, has a low but measurable marginal cost of provision even with methods such as web page data download sites. It is clear that modern data access methods have significantly lowered the marginal cost of provision and, thus, made streamflow data and analysis more clearly a public good. In addition, Internet links and data programs have raised the cost of exclusion, further reinforcing this trend.

The optimal level of provision of streamflow data requires that public recipients reveal the benefits that they receive and that they be taxed in proportion to them. Clearly this process requires a series of "revelation mechanisms" in which a public center receives information from consumers of a public good by providing incentives for its clients to reveal information on the value of the goods; this is necessary to set efficient production levels for the information. One such mechanism is to persuade clients to establish a cost-sharing agreement for location-specific services such as flood warning systems.

In the case of streamflow information, technology can either expand or restrict access to that information. It may not, however, be possible to provide streamflow information to everyone because the cost could not be recovered by those producing the data (such as cooperating non-federal agencies).

Equity Versus Efficiency

Public goods (e.g., the prevention of communicable diseases, the provision of sanitary water supplies) often serve societal values and preferences that motivate their production and supply based on considerations of equity, as well as economic efficiency. Market inefficiencies and market failure associated with public goods may result in distributional impacts that are not acceptable to society. The normative aspects of distributional outcomes reflect value judgments and competing interests that society resolves through the political process rather than market-driven outcomes.

The value of streamflow information may be realized and quantified in, for example, improved infrastructure design (e.g., cost-effectively sizing culverts and bridges). However, the value of this information at the time of design will be very sensitive to the period of record for which information is available. Consequently many of the future benefits and beneficiaries of streamflow information are not fully reflected in current market demand. Current individual pricing and consumption decisions in the competitive market fail to capture the future benefits of current period investment. This further motivates public investment to correct intertemporal market failure.

SUMMARY

The USGS has a history of streamgaging that spans well over a century. Streamflow information supports innumerable planning, management, and scientific activities over a broad range of spatial and temporal scales. These include optimizing hydropower and water supply; reducing impacts of flooding; reducing impacts of droughts; reducing pollutant loads to water bodies; and providing for national defense, food and fiber production, recreation, and wildlife habitat and diversity, including Endangered Species Act requirements. For many specialized applications, the value of streamflow information is enhanced by the density of the streamflow network—that is, the whole is greater than the sum of its parts. In many applications, the direct value of streamflow can be monetized. However, streamflow information displays many of the attributes of the broad class of public goods that are not allocated efficiently through price signals between producers and consumers in competitive markets. This strongly motivates public investment to fully meet the nation's current and emerging needs for streamflow information.

3
Selection of NSIP Base Gage Locations

Chapter 2 showed that the U.S. Geological Survey (USGS) has a long history of gaging streams to meet national needs such as water supply, food supply, power supply, public safety, defense, and many others. Due to finite resources, the USGS has had to prioritize these many needs. This chapter discusses the criteria that the USGS used for locating National Streamflow Information Program (NSIP) base gaging sites to meet what it believed were the five most important national needs, or goals:

1. Meeting Legal and Treaty Obligations on Interstate and International Waters (to monitor legal requirements for deliveries of water at state and national borders; presently 515 gage sites according to *http://- water.usgs.gov/nsip/nsipmaps/federalgoals.html*)

2. Flow Forecasting (sites needed for validation and improvement of forecasts where the National Weather Service and other federal agencies carry out flood or water supply forecasts; 3,244 gage sites)

3. Measuring River Basin Outflows (for calculating regional water balances over the nation; 450 gage sites)

4. Monitoring Sentinel Watersheds (for determining long-term trends in streamflow across the country; 874 gage sites)

5. Measuring Flow for Water Quality Needs (for characterizing the quality of surface waters; 210 gage sites)

A total of 5,293 gage sites are listed under the five criteria but some gage sites serve more than one criterion, so the actual number of gage sites presently identified as NSIP base gages is 4,424 (Figure 3-1). This NSIP base gage network is proposed to be funded 100 percent by the federal government.

Not shown:
Alaska 131
Hawaii 17
Puerto Rico 23

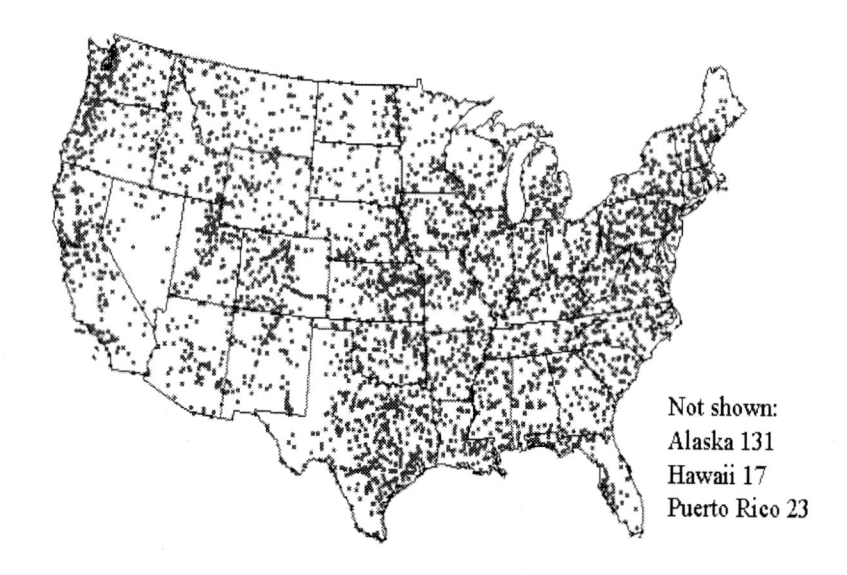

FIGURE 3-1 Locations of the 4,424 gage sites presently identified as NSIP base gages. SOURCE: Based on USGS data (*http://water.usgs.gov/-nsip*).

Of the 4,424 base NSIP gage sites, 2,796 or 63 percent are active USGS gaging locations; 307, or 7 percent, are active gage sites operated by other agencies for which the USGS wants to assume the full costs of operation; 837, or 19 percent, are inactive gages (sites where a gage once operated but no longer does); and 484, or 11 percent, are new gage sites (Figure 3-2). Thus, 3,103, or 70 percent, of all gages presently envisaged for the NSIP are existing gages operated by the USGS or other agencies, and 1,321, or 30 percent, are inactive or proposed new gage sites.

In addition to these 3,103 currently operational gages that comprise the base network and would be 100 percent federally funded, the NSIP includes all of the other currently active, USGS-operated gages. Presently, active USGS-operated gages total about 7,300. Thus, there are many thousands of USGS gages that, although included in NSIP, do not form part of the *base* gage network. This does not mean that these other gages are not fulfilling important purposes, but simply that those purposes may be primarily local in scale or otherwise not of highest national priority as defined by the five federal goals noted above.

Each of the five gage siting criteria is now examined in more detail.

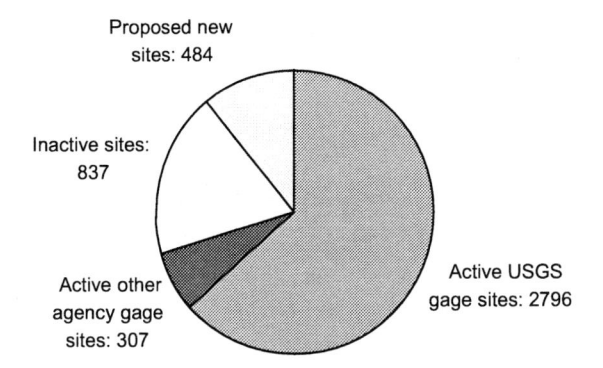

FIGURE 3-2 Status of NSIP gage sites. SOURCE: Based on USGS data (*http://water.usgs.gov/nsip/nsipmaps/usa_sum.html*).

Many additional criteria beyond these five were thoroughly evaluated by the Interstate Council on Water Policy; these are examined in Chapter 4.

THE FIVE CRITERIA FOR SITING NSIP STREAMGAGES

Goal 1. Meeting Legal and Treaty Obligations on Interstate and International Waters

Provide river discharge information to meet the operational requirements of river basin compacts and Supreme Court decrees at each point where major rivers cross international or state boundaries. This goal addresses the need to record the flow of water as an economic commodity across borders and to provide accepted, neutral data for states to use in the allocation of interstate waters.

Metric: Operate a discharge station at rivers

- on or near crossings of state and international borders where the drainage area of the river reach is greater than 500 square miles, or
- where the location is mandated by a treaty, compact, or decree.

A total of 515 gage sites are selected to support this criterion. Of these, 322 serve as border sites, 236 are compact sites, and 43 sites serve both purposes (Figure 3-3). Monitoring streamflow quantity and quality between the United States and adjacent countries is an important mission of the NSIP.

It is prudent to carefully evaluate the status of the ungaged reaches at state boundaries with respect to resource evaluation and environmental needs. Future extraction of water from streams—either direct or induced by enhanced irrigation and other pumping—is difficult to predict, as are future water rights disputes over regional or locally depleting surface water supplies. Where water rights are paramount, as in the Southwest, gaging the volume of water passing by state lines may be increasingly important in the future. The Interstate Council on Water Policy (ICWP, 2002, pp. 6-7) recommended that the NSIP provide streamflow data for rivers governed by compacts between states, tribes, or nations or as dictated under Supreme Court decree but not including waters crossing jurisdictional boundaries with no legal agreements (a summary of all of the ICWP's recommendations is given in Chapter 4). Since water allocation policies and laws differ between states and only states have legal jurisdiction over water originating within the state, it is important to measure all significant interstate flows even if legal agreements or compacts do not yet exist, in anticipation of the data being required for adjudication of future interstate water allocation questions. In contrast to the view of the ICWP, the committee believes that the border gage sites proposed by the USGS should be retained as part of the NSIP base gage network.

Goal 2. Flow Forecasting

Provide real-time data for each of the service locations at which the National Weather Service (NWS) and Natural Resources Conservation Service (NRCS) need streamflow data to calibrate and operate forecast models. Service (NRCS) need streamflow data to calibrate and operate forecast models.

Metric: Operate a streamgaging station at each NWS and NRCS service location that is not located on a reservoir (reservoir locations were excluded because it was presumed that they record water level alone and not discharge).

Selection of NSIP Base Gage Locations 51

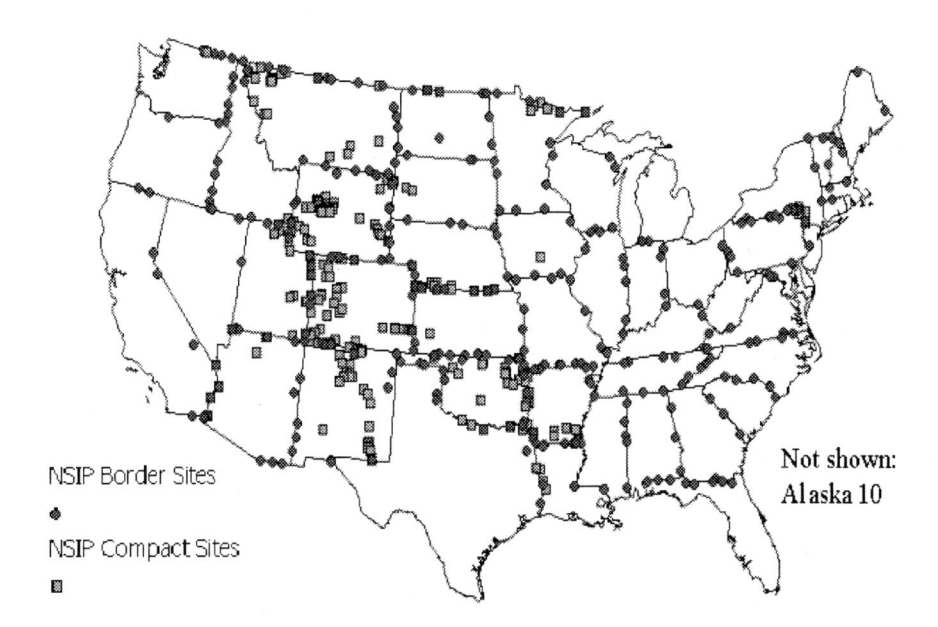

FIGURE 3-3 NSIP border and compact sites (515 gages of which 36 percent are solely for border and compact sites and 64 percent also meet other NSIP goals). SOURCE: Based on USGS data (*http://water.usgs.gov/-nsip*).

The NSIP goal was initially stated only in terms of supporting the NWS flood forecasting program. During the course of the study, the USGS requested that the committee title this goal "Flow Forecasting" rather than "Flood Forecasting" in order to be more inclusive of other needs, such as water supply forecasts, navigation, agriculture, recreation, and drought response. This change is appropriate. **In meeting this broader goal the USGS should include appropriate NRCS gages used in support of water supply forecasting.** The locations of the gage sites identified under this goal (without NRCS gages) are shown in Figure 3-4. Currently, 3,244 gage sites are included in the NSIP by the flow forecasting criterion, which is 76 percent of the total number of gages in the NSIP (4,424).

The principal sources of gage sites for the NSIP are the National Weather Service's river forecasting points (Box 3-1). In many cases, these locations were originally determined because a USGS gage existed there. In other cases, the forecast point locations are determined by local needs for

Not shown:
Alaska 10
Hawaii 5
Puerto Rico 16

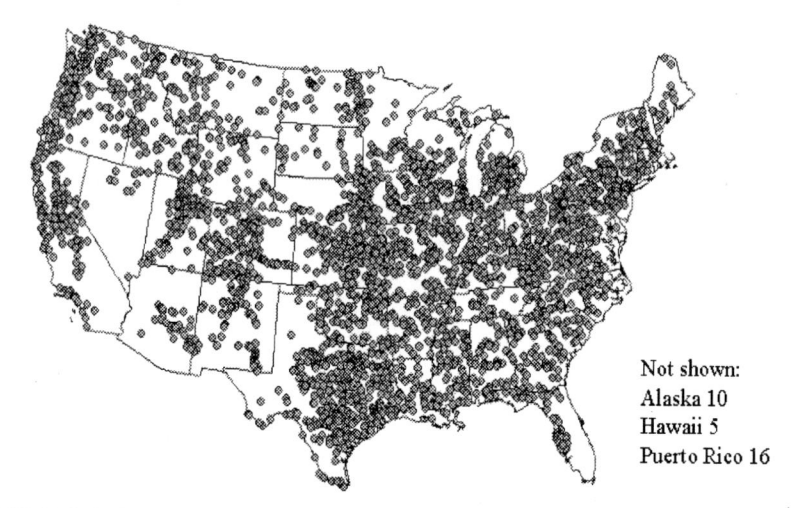

FIGURE 3-4 NSIP gage sites for flow forecasting (3,244 gages of which 80 percent are solely for flow forecasting and 20 percent also serve other NSIP goals). SOURCE: Based on USGS data (http://water.usgs.gov/nsip).

forecast information or by a reasonable subdivision of the landscape into forecast watersheds. Decisions about these forecast point locations are made in the 13 River Forecast Centers. The NRCS also carries out long-term (a few months in advance) forecasts for water supply in the western states using remote sensing and ground-based measurements of snow as an input variable.

Figure 3-5 shows NRCS water supply forecast gage sites. These locations are all USGS gage locations in the western United States where water supply forecast needs are most critical. Of the 576 NRCS gage sites, 321 are already in the NSIP base gage network, so the addition of the NRCS sites would add 255 sites to the 3,244 NWS sites currently identified for the flow forecasting goal, an increase of 8 percent. **This is a modest increase in the total number of sites in this category and is well justified in support of the broader goal of flow forecasting, as distinct from flood forecasting.**

The USGS streamgaging network is an integral part of the forecasting process. Mason and Weiger (1995) reported that in 1995, the NWS utilized real-time data from 3,971 USGS streamgages to make forecasts at 4,017 forecast points (the total number of forecast points at that time). Furthermore, USGS accounted for 98 percent of the stations used by the NWS for

BOX 3-1
River Forecasting at the National Weather Service

Forecasts of river conditions provide vital information for flood warning, water management, navigation, and recreation. Although many federal, state, and local agencies engage in streamflow forecasting to meet operational objectives, the NWS is the agency responsible for issuing river forecasts and flood warnings to the public, as mandated by the National Weather Bureau Organic Act of 1890 (U.S. Code title 15, section 311).

River forecasts issued by the NWS provide site-specific information on river conditions at more than 4,000 and "forecast points" across the country. The nature of the forecasts can vary, depending on the concerns at a forecast point. At some locations, forecasts are issued on a daily basis. The forecasts often consist of river stage hydrographs (water level versus time) for the next few days. At other locations, forecasts are issued only during floods. These flood-warning forecasts consist of the expected time and river stage of the flood crest, as well as the period during which the river stage is expected to be above flood stage.

The numerical guidance used to issue the forecasts is produced at one of 13 NWS River Forecast Centers (RFCs). However, the forecasts themselves are issued by a service hydrologist at one of the 121 NWS Weather Forecast Offices (WFOs). The WFOs can also issue general flood watches and warnings for areas within their region. When weather conditions warrant, flood warnings are issued for rivers and streams within designated areas where flooding may occur (rather than at specific forecast points). The guidance that hydrologists use to issue flood warnings can include forecast products generated at the RFCs or products generated locally at the WFOs.

The forecasting techniques used by the NWS have evolved over the years. Early forecasting techniques combined observations of river and weather conditions, simple hydraulic and hydrologic methods, and a forecaster's experience and judgment. Today's forecasting technologies are a product of recent advances in computer power and data telemetry. For instance, real-time data on river discharge and stages, precipitation from raingages and NEXRAD weather radars, and other weather observations, are transmitted via satellite for access by the RFCs. These data are then fed into computer models that simulate watershed processes and river hydraulics. The suite of hydrologic and hydraulic models used in river forecasting at the RFCs is known collectively as the NWS River Forecasting System (NWSRFS). In the past decade, NWS has begun using forecasts of future precipitation (known as quantitative precipitation forecasts) to improve short-range streamflow forecasts.

Hydrologists at the RFCs produce the numerical guidance for issuing river forecasts and flood warnings by running NWSRFS models for river basins within their area. A river basin is represented in an NWSRFS model as a set of subcatchments or subbasins. The hydrologic response of the subcatchment to rainfall and snowmelt is simulated with observed weather and streamflow data and forecast information. Streamflow is predicted at the subcatchment outlet. To ensure that river forecast guidance is available at each forecast point, each forecast point corresponds to one of the subcatchment outlets. The remaining outlets (which are not forecast points) are known as *data points*. The passage of upstream outlet flows through downstream subcatchments is simulated using hydraulic routing models. Where river flows enter a reservoir, future reservoir releases are needed to forecast downstream flows. In some locations, the NWS contacts the agencies responsible for reservoir operations to obtain this information. In other locations, the NWS sends its reservoir inflow forecasts directly to the operational agency. These agencies then utilize the streamflow forecasts in their operational model to plan future releases and report their plans back to the NWS.

As part of the modernization of weather and hydrologic services, the NWS has begun implementation of Advanced Hydrologic Predictions Services (AHPS). These services include new forecast products and web-based visual displays of forecast information at forecast points. Some of the new products are long-range probabilistic streamflow forecasts. Probabilistic forecasts of river stages and discharges, flood stages and discharges, and river flow volumes are made or 90 days. Examples of the AHPS products are available on-line for portions of the Upper Mississippi River (*http://www.crh.noaa.gov/dmx/ahps/*) and the Ohio River (*http://www.erh.noaa.gov/er/ohrfc/ahps.htm*).

In addition to river forecasting and flood warning, the NWS also issues flash flood watches and warnings. Unlike river forecasts, flash flood forecasts are issued for counties or areas rather than specific forecast points. The forecasting process often involves real-time comparisons of observed rainfall from gages or NEXRAD weather radar with estimated flood-producing rainfall amounts, known as flash flood guidance. The flash flood guidance depends on the soil moisture state of the drainage area and is updated daily by the RFCs with information from the NWSRFS models used in river forecasting. Another approach to flash flood warning is the NWS Integrated Flood Observing and Warning System (IFLOWS) in the Appalachian Region. This system continuously monitors rainfall and water levels on streams, and automatically sends out warnings to emergency managers. Similar systems, called Automated Local Evaluation in Real-Time (ALERT) systems, have been implemented by state and local agencies for flash flood warning in other parts of the United States.

real time river observations. River forecasting is a data-driven process, and streamflow information is the most important data source. NWS forecast points and USGS streamgages are collocated so that measured streamflow data can be used to calibrate and initialize forecast models. Accurate and reliable forecasts require both real-time streamflow information for model initialization and a historical record of streamflow information over many years (or decades) for calibration (streamgaging data in catchments under going land-use change can also be used to update model parameterization). Often, knowing the river flows at upstream locations is the most critical component in making an accurate forecast at downstream forecast points.

Selection of NSIP Base Gage Locations 55

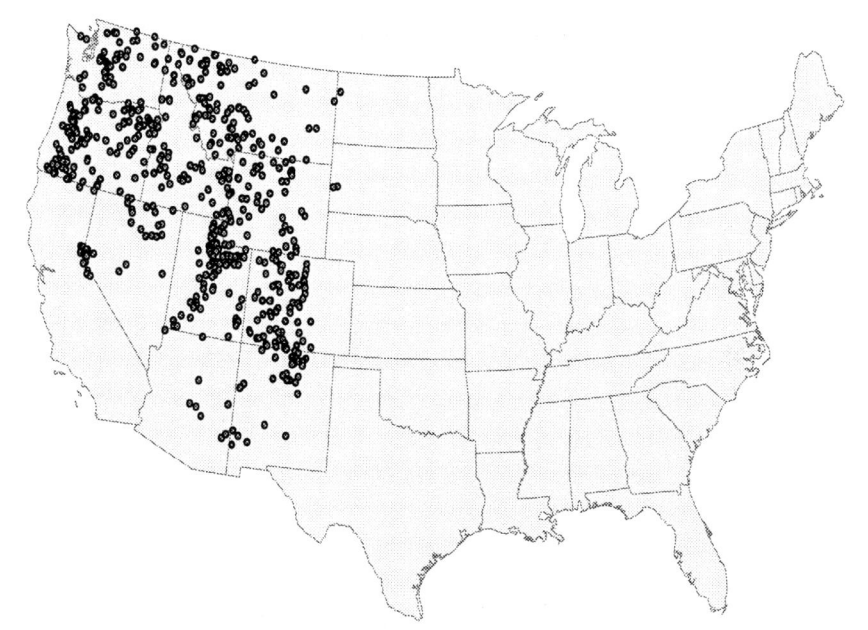

FIGURE 3-5 NRCS gage sites used for water supply forecasting (576 gages). SOURCE: Data provided by David Stewart, USGS, written communication, 2003).

Observations are also used by service hydrologists to adjust model predictions in real time to make better river forecasts. Despite the complexity of forecast models, model predictions diverge from actual conditions due to uncertainties in the measurement of precipitation and other weather input variables or the inherent simplifications associated with computational modeling of watershed and river processes. Real-time observations provide the "ground-truth" needed to continuously make reliable forecasts for rapidly changing river conditions. Even though improvements in hydrologic forecast models are expected in the future, real-time observations will always be necessary for model calibration.

Historical information is also essential for development of accurate streamflow prediction models. Due to the complex processes controlling runoff and river flows, it is not possible to accurately predict river flows without first comparing model predictions to observations and then making adjustments to the model parameters to improve predictions—a process known as calibration. In operational forecasting, flow extremes (both floods and drought conditions) are generally when the need for accurate forecasts is most critical. Flow extremes are also rare events. As a result, an extended record with both dry and wet extremes is necessary for model calibration. Additional flow information (not used in model calibration) is also needed to evaluate the model's predictive ability.

In times of flood, the public is most interested in the height to which the river will rise, not its discharge rate. At each USGS streamgage station, hydrologists develop and maintain a rating curve, which relates discharge to river stage (and vice versa). However, rating curves are dynamic, changing with changes in channel shape (which often occur from the erosion and deposition associated with floods). The NWS includes these rating curves directly in its forecast models to make river stage forecasts.

Since USGS gage sites were selected as forecast points early on in the river forecasting process at the NWS, the hydrologic models used by the NWS have been designed around the long-term streamgage network. For example, the delineation of model elements (subcatchments) is driven in large part by the location of long-term gages. Matching the outlets of model subcatchments to locations with long-term streamflow records facilitates the forecast model calibration step. From the perspective of an operational forecaster, the historical streamflow record at a site can do more to improve forecasting than new observations from a previously ungaged site. This occurs because a revised forecast model can be reliably calibrated to make predictions at a point with historical information. Hence, given a choice, many operational forecasters would choose to restore discontinued gages (for their historical record) rather than establish new streamgage sites.

Continued operation of long-term streamgaging stations is also vital to the NWS river forecasting mission. A chief concern for operational forecasters is the loss of existing streamgages. Although a forecast model can continue to make predictions at the site, the loss of real-time observations needed to adjust model predictions degrades the accuracy of forecasts. As time passes, rating curves will shift, further reducing forecast quality. As a matter of procedure, forecast points are not eliminated when a streamgage is discontinued, but service hydrologists are instructed that forecast guidance is to be interpreted qualitatively, rather than quantitatively, in issuing forecasts (NWS, 2003).

Goal 3. Measuring River Basin Outflows

Provide representative data for each of the major river basins in the nation.

Selection of NSIP Base Gage Locations 57

Metric: Operate streamgaging stations near the terminus of each of the 352 hydrologic accounting units (also called six-digit hydrologic unit code basins, or HUC-6) in the nation (see Figure 3-6). The intent of the metric is to gage as much of each unit as possible. For accounting units drained by a single major river, a streamgage should be located near the outlet of the accounting unit so that the drainage area to the streamgage is 90 to 110 percent of the accounting unit drainage area. For a coastal unit or area of internal drainage, the farthest downstream station or gage location on the largest river is selected, and if this location drains less than 50 percent of the accounting unit area, then the location with the next-largest drainage area on a different river within the accounting unit is selected.

One of the key goals of water resources planning for the United States is to be able to perform a water balance of the flow of water through the nation. Such a balance, in which a budget is constructed of all the inputs and outputs of water in each river basin or water planning region, is required for a myriad of applications. These include national assessments of water availability and water use and their change over time, effects of major changes in basin management (e.g., dam construction or removal) or water supply (e.g., conversion from groundwater to surface water or vice versa) on streamflow and biota, and others. To support such an accounting the USGS has identified the six-digit hydrologic cataloging units as the regional water planning units that such a water balance will require and has located an NSIP gage site near the outlet of each of these six-digit hydrologic regions. A total of 450 gages are located by this criterion (Figure 3-6).

Goal 4. Monitoring Sentinel Watersheds

Provide data from stations that are minimally affected by human activity for regionalization of streamflow characteristics and assessments of trends in streamflow due to factors such as changes in climate, land use, and water use.

Metric: For the conterminous United States, a set of 802 eco-accounting units was defined by intersecting the 352 accounting units with the 76 ecoregions of United States defined by Omernik (1987) and selecting all the resulting areas greater than 100 square miles. The result is presented in Figure 3-7.

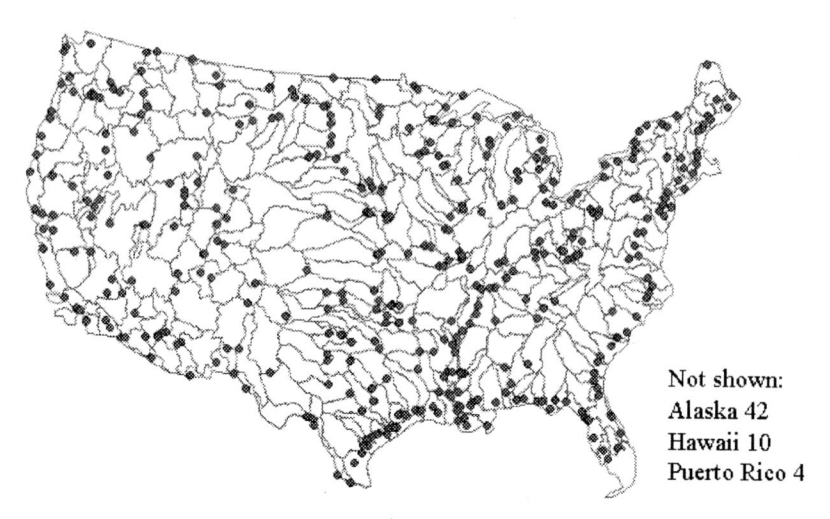

58	*Assessing the National Streamflow Information Program*

Not shown:
Alaska 42
Hawaii 10
Puerto Rico 4

FIGURE 3-6 Gage locations at river basin outflows (450 gage sites of which 36 percent are solely for river basin outflows and 64 percent also serve other purposes). Lines shown are boundaries of hydrologic accounting units. SOURCE: Based on USGS data (*http://water.usgs.gov/nsip*).

Not shown:
Alaska 20
Hawaii 9
Puerto Rico 5

FIGURE 3-7 Gage locations for sentinel watersheds (874 gage sites of which 71 percent are solely for sentinel watersheds and 29 percent also serve other purposes). Lines shown are boundaries of ecoregions. SOURCE: Based on USGS data (*http://water.usgs.gov/nsip*).

Gage sites were selected to be relatively free from human influence (no large reservoirs upstream) and have 80 percent of their drainage area within a single ecohydrologic unit. For Alaska, one station was selected for each ecological region rather than for each ecohydrologic region because the cost of operating a gage station in Alaska is about $50,000 a year, as compared to $10,000 per year in the conterminous United States. For Hawaii and Puerto Rico, a windward and leeward station was selected for each major island. Preference was given to selection of stations in the Hydroclimatic Data Network (HCDN) (Slack et al., 1993) where possible, because the records at these stations have been carefully checked for accuracy and consistency.

One of the key goals of a national streamflow program is to monitor long-term trends of streamflow in the nation. This is particularly important in relation to climate change, which could influence the frequency and severity of floods and droughts. To meet this criterion, the USGS has constructed a set of hydro-eco regions by intersecting the Omernik ecological regions, adopted by the U.S. Environmental Protection Agency (EPA) as a good description of ecological variation over the United States, with the six-digit HUC basins and has identified a nearby gage site where flow is minimally impacted by upstream storage or diversions. A total of 874 gage sites are identified in the NSIP by this process. These gages are viewed as critical for developing the emerging river science program (see Chapter 6). One consideration in evaluating the NSIP base gage network and sentinel gage subset is the coverage in terms of spanning a representative range of basin sizes. This is illustrated in Figure 3-8, which shows, in aggregate, that sentinel gages are well distributed over drainage areas and reasonably that both the sentinel gaging and the overall USGS gaging networks have a reasonable distribution of basin sizes, with the sentinel gages placing slightly more emphasis on smaller watersheds.

Goal 5. Measuring Flow for Water Quality Needs

Provide streamflow information for a national network of water quality (concentration and loading) monitoring points.

Metric: There are three national water quality networks: Hydrologic Benchmark (HBM) (63 stations), National Stream Water Quality Accounting Network (NASQAN) (40 stations), and National Water Quality Assessment Low-Intensity Phase (NAWQA-LIP) (107 stations). Active streamgaging stations were required to be on the same river reach as the water quality monitoring site. The location of these gages is shown in Figure 3-9.

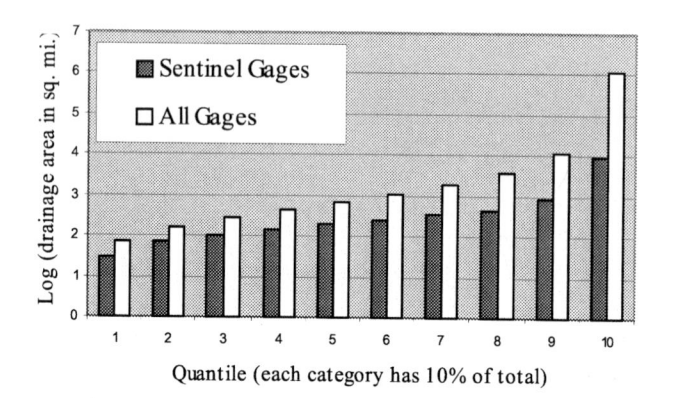

FIGURE 3-8 Drainage area represented by all USGS gages and by sentinel gages. Explanation: 10 percent of all USGS gages monitor drainage areas of 10^4-10^6 (10,000-1,000,000) square miles, 10 percent monitor areas of $10^{3.5}$-10^4 (3,200-10,000) square miles, and so on. The distribution shows that sentinel gages are well distributed over drainage areas and reasonably reflect the overall gage distribution. SOURCE: Data provided by J. Michael Norris, USGS, personal communication, June 2003.

FIGURE 3-9 Gage locations for water quality (210 gage sites of which 58 percent are solely for water quality and 42 percent also serve other purposes). SOURCE: Based on USGS data (*http://water.usgs.gov/nsip*).

Proper interpretation of water quality data requires knowledge of stream discharge. Gages for this coverage were selected primarily to meet NAWQA needs. In the past, funds from the NASQAN program also supported a small number of streamgages, and in some cases funding support for these gages has been transferred to the NSIP. The NASQAN program itself has been reduced in scope in recent years as the focus on water quality assessment has shifted to the NAWQA program. The HBN stations are part of the Hydroclimatic Data Network set of gages and are already receiving federal support independently of the cycles of funding for the NASQAN and NAWQA programs.

The NSIP also supports other water quality needs. For example, EPA's Total Maximum Daily Load (TMDL) program requires estimates of flow to determine chemical loads and transport. According to the USGS, based on concentrations of a range of water quality pollution indicators, 677 out of a total of 2,079 possible watersheds in the conterminous United States were identified as degraded. Of these degraded watersheds, 85 percent are adequately gaged, which indicates that NSIP gaging is supporting water quality needs beyond the minimal set of sites designated under this specific goal.

However, additional gaging to support the TMDL program on a site-specific basis would be overwhelming in cost and manpower because there are about 21,000 polluted river segments, lakes, and estuaries making up more than 300,000 river and shore miles and 5 million lake acres (NRC, 2001). There is a pressing need to be able to estimate historical and real-time streamflow at any point on a river network. This goal can be met only by new scientific research, based on existing streamflow data, to develop accurate regionalization methods. The USGS is well positioned in terms of expertise to do this research.

ASSESSMENT OF THE DISTRIBUTION OF GAGE SITE LOCATIONS

The selection of gage sites using the five NSIP criteria reflects a process of assessment within the USGS as to which goals out of all those used to justify installing a gage site are appropriately national or federal goals and which goals are better left for state and local interests. Among the five criteria, the flow forecasting criterion to support the river forecast operations of the NWS results in many more site locations than any other criterion (Figure 3-10). Of the 5,293 sites selected, 3,244, or 61 percent, support the forecast goal; 874, or 16 percent, support the sentinel watershed goal; 515, or 10 percent, support the border or compact goal; 450, or 9 percent, sup-

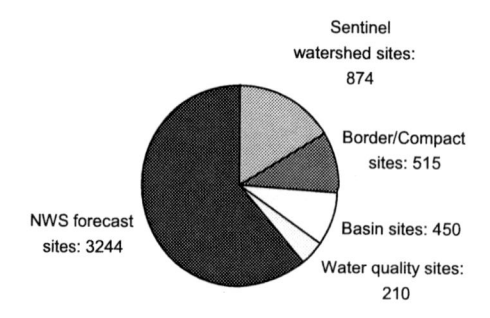

```
                    Sentinel
                 watershed sites:
                       874

                                    Border/Compact
                                      sites: 515

    NWS forecast
     sites: 3244                    Basin sites: 450

                                  Water quality sites:
                                          210
```

FIGURE 3-10 Distribution of NSIP gage site locations by goal. SOURCE: Based on USGS data (*http://water.usgs.gov/nsip*).

port the basin accounting goal; and 210, or 4 percent, support the water quality goal. Of course, some site locations support more than one goal; only 4,424 site locations are needed to support the 5,293 locations selected independently by the five criteria.

One way of assessing the result of this selection process is to examine the distribution of gage sites across the country. Figure 3-11 shows the distribution of gages by state, with the largest numbers of gages being located in Texas (416), followed by California (201), Colorado (171), Kansas (166), Oregon (136), and Alaska (131). The states or commonwealths with the smallest number of NSIP site locations are Rhode Island (2), Delaware (4), Vermont (15), Connecticut (16), Puerto Rico (17), and Maryland (18).

A key criterion typically used to measure the adequacy of streamgage networks is gage density, measured as the number of gages per square kilometer of land area. Figure 3-12 shows the gage density in the 48 conterminous states. The states with the greatest gage site density are Puerto Rico, New Jersey, Massachusetts, Connecticut, Pennsylvania, and Hawaii, which all have more than one gage per 1000 km^2. The states with the lowest gage density are Alaska, Nevada, South Dakota, Arizona, Maine, New Mexico, Montana, and North Dakota, which all have less than one gage per 2500 km^2.

There is a significant discrepancy between the lowest two states in gage density—Alaska with one gage per 11,700 km^2 and Nevada with one gage per 9650 km^2—and the next four states—Arizona, South Dakota, Maine, and New Mexico, which each have about one gage per 3000-3500 km^2.

Selection of NSIP Base Gage Locations 63

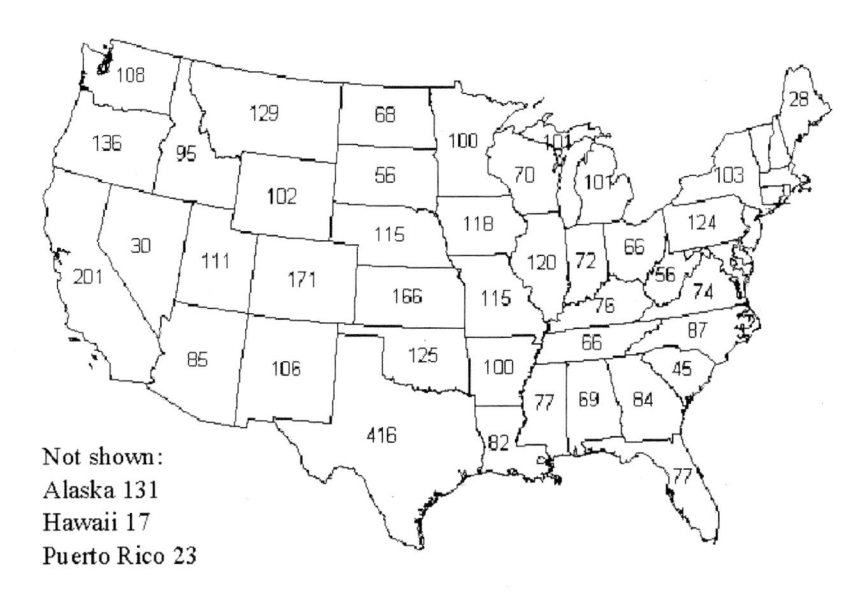

FIGURE 3-11 Number of NSIP gage sites by state. Totals for northeastern states: Connecticut, 16; Delaware, 4; Maryland, 18; Massachusetts, 23; New Hampshire, 20; New Jersey, 29; Rhode Island, 2; and Vermont, 15. SOURCE: Based on USGS data (*http://water.usgs.gov/nsip*).

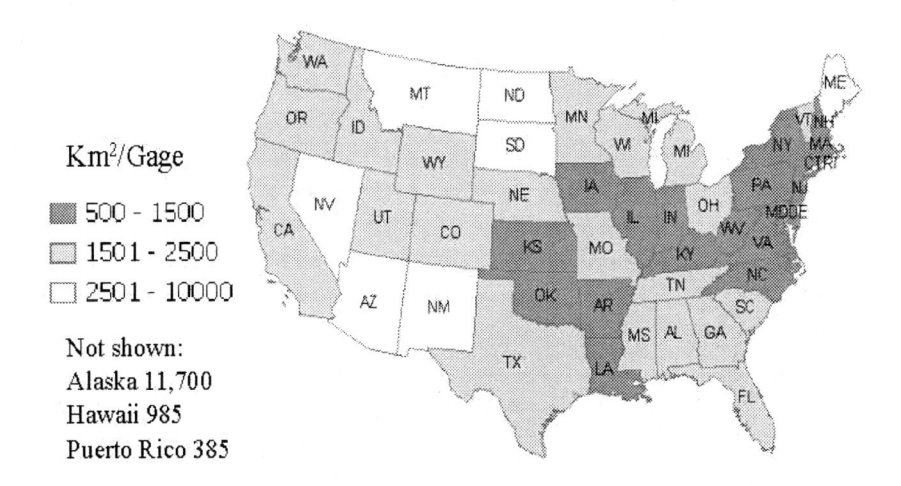

FIGURE 3-12 Density of NSIP gage locations measured in square kilometers of land area per gage. SOURCE: Based on USGS data (*http://water.usgs.gov/nsip*).

Alaska is a special case because of the very large land area of the state and the high cost of operating gaging stations there, but the low density of gages in Nevada cannot be justified in that way. Figure 3-11 shows that there are only 30 NSIP gage site locations in Nevada, while its neighboring states—Arizona, Utah, Idaho, and Oregon—have 85, 111, 95, and 136 sites, respectively. Apart from Arizona, which is slightly larger, these neighboring states are all smaller than Nevada.

There are several reasons for the anomalously low gage density in Nevada. One contributor is that in the interstate boundary category, most of the gages near the Nevada border happen to be in the adjacent state. Another is that there are only 10 National Weather Service river forecast points in Nevada, compared to 34 in Arizona, 89 in Utah, 62 in Idaho, and 109 in Oregon. This variation in NWS forecast points among the states closely matches the variation in NSIP gage distribution in those states and reaffirms the importance of the existence of NWS forecast points in siting NSIP gages. The low number of NWS river forecast points is due in part to the nature of Nevada's hydrology. Streamgaging is difficult and ineffective for flood forecasting in ephemeral streams. In particular, the NWS continuous simulation model time step is too long to be effective for the rapid forecasts necessary for flood forecasting in ephemeral streams. In such circumstances, raingages are more effective for flood warning, exemplified by Las Vegas' local alert service.

There are 28 NRCS water supply forecast sites in Nevada, and of these, 11 are already in the NSIP base gage network. Thus, the addition of the NRCS forecast points to the NSIP base gage network would add 17 sites to the 30 currently in network in Nevada, making a total of 47 sites. Therefore, adding the NRCS forecast sites would not eliminate the discrepancy for Nevada, but it would ameliorate that discrepancy somewhat.

Another reasonable way of assessing network adequacy is to determine the number of gage sites per person in each state (Figure 3-13). While this is not a standard method, public safety, in the form of the supporting NWS flood forecasting, generates far more NSIP sites than any other criterion. It follows that if public safety is implicitly a goal, then more measurement may be needed where more people are threatened by floods. Alaska has the lowest number of people per gage (4,200), while Rhode Island (494,000), Maryland (289,000), New Jersey (282,000), and Massachusetts (269,000) have the highest. Rhode Island's anomalously high ratio is probably not significant; there only two NSIP gage sites in Rhode Island and the addition of only two more sites there would bring it into line with the other states just listed. It can be seen by comparing Figures 3-12 and 3-13 that although Montana, North and South Dakota, Maine and New Mexico all have a low

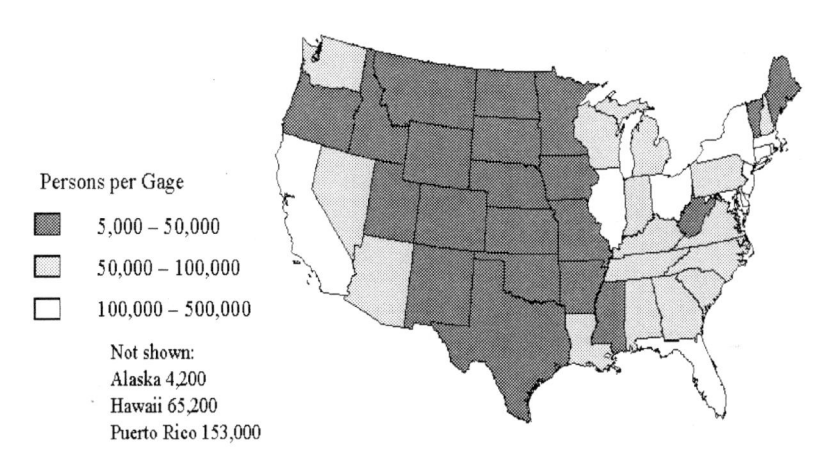

FIGURE 3-13 Number of persons per NSIP gage site in each state. SOURCE: Based on USGS data (http://water.usgs.gov/nsip).

gage density per unit area, these states all have a high gage density per person, so the number of NSIP gage sites assigned to these states seems reasonable. Nevada has 60,600 persons per gage, which ranks about in the middle of the states in this criterion.

These examples highlight not only a specific shortcoming of the NSIP with respect to ephemeral streams, but the general principle that certain locations may have greater value in the future than is presently perceived. The principle of adaptive management should be incorporated explicitly into the NSIP to periodically reevaluate the network goals and criteria to ensure that the network meets present and anticipated future needs for streamflow information. This periodic reassessment of emerging needs may also support gaging of streams in small watersheds or coastal plains where there is a perception of insufficient coverage and common sense dictates the inclusion of gage sites that may not be included by rigidly applying the five current criteria.

In addition, the USGS should consider how the public, the scientific community, and water management agencies will be included in the adaptive management of this national network. At present, much of the public input on prioritizing streamflow gaging comes in the form of having paying state and local customers through the Cooperative Water (Coop) Program. If the NSIP fully funds its base network independent of cost matching, other mechanisms for public consultation at various levels (e.g., an advisory board, surveys) will have to be found.

SUMMARY

The five criteria that the USGS has used to prioritize gages for the NSIP seem reasonable. The distribution of gages by state across the nation produced by the NSIP criteria also appears reasonable when measured on metrics of number of gages per unit of land area and number of persons per gage, with the possible exception of Nevada, which has only 30 NSIP base gage sites. Nevada has about one-third the number of NSIP base gage sites of its neighbor states, due to the National Weather Service having only 10 flow forecast points in Nevada. Bearing in mind the rapid growth of Nevada and its critical dependence on water resources, reexamination of the number of NSIP gage sites there may be warranted. The overall distribution of sentinel gages with respect to watershed size is also reasonable.

The principle of adaptive management should be incorporated explicitly into the NSIP. This periodic reevaluation of network goals and criteria will ensure that future needs for streamflow information, including ephemeral streams and possibly small watersheds and coastal plains, are met by the network.

The NSIP gage site program is very much attuned to the site locations of National Weather Service river forecast points, and vice versa. Although there are five gage siting criteria, more than 60 percent of the sites selected for NSIP gages are determined by the locations of NWS forecast points. The USGS deals with streamflow information in the past and the present. The National Weather Service is responsible for taking current streamflow information and forecasting future flows. It appears that in the past there has been limited coordination between the USGS and flow forecasting agencies (NWS and NRCS) in the location of gages to meet the flow forecasting goal. The NWS opportunistically locates a forecast point where there is a USGS streamgage. The USGS justifies the presence of the majority of the NSIP base streamgages as supporting flow forecasting by the NWS or NRCS.

This raises the question of whether the flow forecasting gages are optimally located to support forecasting needs. Decisions regarding the location of flow forecast points are made regionally within the 13 River Forecast Centers and NRCS offices responsible for water supply forecasts. It appears that a greater degree of cohesion between the USGS and the NWS in planning and locating future gage sites and forecast points would be beneficial, especially in western states such as Nevada. New NSIP gages are sited in consultation with USGS district offices, which in turn are charged with taking account of user needs in their districts. In addition, **a formal**

coordination mechanism should be established between the NWS, NRCS, and USGS for the selection of flow forecast NSIP base gages. This coordination should consider the national NSIP coverage model proposed in the following chapter.

4
Streamflow Network Design

The question of where to site streamgages and how long to maintain them at these sites is a central one for hydrologic data collection agencies throughout the world. Many approaches have been used to design and maintain data collection networks. In the past, network design approaches at the U.S. Geological (USGS) and elsewhere have relied largely on statistical methods, most commonly based on the standard error in estimating regional discharge at ungaged sites. Although statistical procedures offer numerical precision for network design supporting regional hydrologic estimation, these approaches do not support the many other goals and uses of site-specific streamflow data. In contrast, coverage models are based on articulating a goal, defining a measure of success ("metric") or procedure that identifies locations supporting that goal, and applying this procedure using geographic information system (GIS) analysis to yield a set of potential sites (e.g., for gages). One advantage of this approach is that it yields discrete yes or no answers about site locations for each goal considered.

This chapter considers and contrasts these approaches to network design and maintenance. The proposed National Streamflow Information Program (NSIP) gage network is considered in this broader context, including comparisons with state-level hydrologic networks, and the evaluation of other reviews of the NSIP. The chapter concludes with a vision of the NSIP as a national information program with the broad goal of providing streamflow information with confidence limits at any arbitrary point in the landscape.

STATISTICAL MODELS

The most common network design methods have been based on statistics. During the 1970s and 1980s, the USGS developed and applied statistical regression techniques to locate gages (Moss, 1982; Stedinger and Tasker, 1985; Tasker, 1986). More recently, other investigators have used entropy methods and other statistical concepts to quantify relative information content (Bueso et al., 1998; Lee, 1998; Mogheir and Singh, 2002; Perez-Abreu and Rodriguez, 1996). These studies invoked the strong assumption that streamflow observations (and therefore climate and land use) are stationary. For the narrow, well-defined problems of hydrologic regionalization and the estimation of specific flow quantiles (such as the 100-year flood) at ungaged sites, the information content of additional streamflow observations can be quantified by the decreasing standard error of the estimate.

Network Design

Considerable research has been done on the design of monitoring networks in the earth sciences. Perhaps most common are networks designed to use observations at discrete points in space and time to estimate the characteristics of a continuous field or flux (Bastin et al., 1984; Bras and Rodriguez-Iturbe, 1976; Pardo-Igúzquiza, 1998; Rodriguez-Iturbe and Megia, 1974; Sampson and Guttorp, 1992). If the spatial and temporal structure of the variable of interest (e.g., precipitation, evaporation) is well known, its value at any arbitrary location within the network can be estimated using this approach (Boer et al., 2002; Zidek et al., 2000). Geophysical networks can similarly be designed to estimate the position and magnitude of seismic events (Havskov et al., 1992) or to optimize the sensitivity and probability with which movements of the earth's crust can be detected. In contrast to these networks, streamgages are located only on the streams themselves, rather than throughout the entire catchment. Measurement nodes in the stream network provide estimates of fluxes or concentrations of particulate and dissolved constituents. Streamgage networks may be driven by the need for information at a specific location, such as concentrations or fluxes where a river enters a waterbody or crosses an in-ternational boundary, or a critical flood warning site. For these needs the gage site is fixed.

For other applications, the site at which streamflow information is needed is characterized only by the properties of the contributing upstream drainage area. For example, evaluation of hydrologic and ecological effects of land conversion from forest to agricultural use requires streamflow measurements from watersheds experiencing these land-use changes. Many different candidate sites can satisfy this type of information need, expanding the flexibility (and complexity) in designing a streamgage network. An important class of management problems requires streamflow data from gages that sample "representative" locations, to support regional modeling, estimation, and trend detection.

Information theory offers a formal approach to network design by quantifying the marginal contribution of each data collection node to the overall information provided by a network. This incremental value can be formally measured in probability terms by the "cross-entropy" of an event on the preexisting probabilities. Shannon (1948) showed that a "measure of how much 'choice' is involved in the selection of the event or of how uncertain we are of the outcome" (H) must have (which relates to its information content) would be proportional to -ln (p), where p is the prior probability of the event happening.

Shannon (1948) also extended the definition from single probabilities to discrete distributions and defined the expected information content of a prior distribution $\Sigma_i p_i$ as the entropy of a distribution:

$$H = - \Sigma_i p_i \ln p_i. \qquad (1)$$

It follows that a uniform distribution, in which each event is equally likely, has the highest entropy and the lowest information content. Conversely, a distribution that puts a weight of 1 on a single outcome and zero on the rest has an entropy of zero and the highest information content.

The concept of cross-entropy (CE) as a measure of incremental information gain was extended by Kullback and Liebler (1951) and defined as

$$CE = - \Sigma_i p_i \ln (p_i/q_i); \qquad (2)$$

where q_i is the set of prior probabilities that are held by the decision maker.

The importance of this theory to streamflow information is that it follows from equation (2) that if the new signals (e.g., for stage or discharge) coming from a monitoring system (p_i) (e.g., a gage) are close to those expected from the prior probability (q_i) generated from past streamflow observations, then $\ln (p_i/q_i)$ tends to zero and very little information has been added to the system. The converse, of course, is also true. For more theory, see Cover and Thomas' (1991) *Elements of Information Theory*.

In the narrow context of hydrologic regionalization, quantifying incremental information in this way can support the formulation of a formal network design problem to maximize the trade-off between network information content and network cost. In contrast, the breadth of both the national NSIP goals and the hydroclimatic variation spanned by the NSIP network is not meaningfully reduced to a simple set of statistical measures. Thus, the most appropriate role for these methods for NSIP is supporting the analysis of incremental refinements to local and regional hydrologic networks, within the broader context of the NSIP network design. Within this formality, distinct variations of this decision problem have been described and applied in network reduction, network expansion, and network refinement.

Network Reduction

Commonly, an existing network must be evaluated to determine which gages to discontinue when the network must be reduced (Boer et al. 2002; Oehlert 1996), for example, due to budget cuts. The USGS has abundant experience with this problem. "This network reduction" decision problem involves minimizing the information loss associated with discontinuing gages, subject to a constraint on the number of gages to be discontinued (a surrogate for the total cost reduction that must be achieved). In this case, records from each of the gages in the network provide observational data that can be used to quantify the information loss associated with eliminating each gage based on testable assumptions of regional homogeneity and stationarity. Monte Carlo experiments can be used to rigorously quantify this information loss over specific statistical measures, such as the change in the standard error of regionalized estimates of discharge quantiles (e.g., the 100-year flood).

Network Expansion

The complementary decision problem involves maximizing the information increase associated with adding gages to an existing network. Though similar, the network expansion problem requires an estimate of the information content to be gained from previously ungaged candidate sites. As in the case of network reduction, the accuracy of this estimate depends on understanding how the value of the variable of interest changes as a function of its position in the stream network, location in the landscape, topographical position, and other watershed attributes. The accuracy of this estimate (which determines the performance of the enhanced network) is based on assumptions of regional homogeneity and stationarity (i.e., invariance of the underlying random processes with respect to time) within the network (Haas, 1992). Unlike the case of network reduction, for network expansion this assumption is less easily tested, since observations are obviously not yet available at new gage locations.

Network Refinement

A third variation of the network design problem involves adding new gages to a network when neither the candidate locations nor the number of gages to be added has been decided a priori. Such is commonly the case in designing a network of groundwater monitoring wells where the location of the wells (e.g., relative to the estimated position of a contaminant plume) and the number of wells to be added are both decision variables. This problem similarly requires an indirect estimate of the information contributed by each new well derived from an underlying structural model of the currently unobserved system. For the NSIP network design, candidate gage locations are effectively unlimited.

The most general problem with respect to deciding to remove or add new NSIP gages combines all three decision problems in which an existing monitoring network is to be improved through the combination of adding gages, discontinuing gages, and locating new gages. All of these approaches require continuous, well-defined information metrics that can be expressed as a function of the number and/or location of gages. In well-defined networks with limited objectives, statistical approaches for network design can be used to evaluate incremental decisions to add or eliminate individual gages within a local gage network serving narrow, well-defined goals, such as estimating flows at ungaged sites. An example of such an application is given in the following section.

Example of a Statistical Network Design: Texas

Statistical approaches to design regional streamgage networks are exemplified by a recent study to assess the state streamgage network (Slade et al., 2001) conducted by the Texas District of the USGS and the Texas Water Development Board. The goals for the Texas streamgage network were the following:

- *Regionalization*—estimate flows or flow characteristics at ungaged sites in 11 hydrologic regions of Texas

Streamflow Network Design 73

- *Major flow*—obtain flow rates and volumes in large streams
- *Outflow from the state*—account for streamflow leaving the state
- *Streamflow conditions assessment*—assess current conditions with regard to long-term data and define temporal trends in flow

As shown in Figure 4-1, in 1996, Texas had 329 streamflow of which 312 stations were continuous flow recorders and 17 were peak flow stations. The number of continuous flow recorders reached a maximum of about 420 gages in 1972 and declined thereafter. The NSIP goal for Texas is 416 gages, a number that was actually exceeded for about five years in the 1970s. The downward trend in streamgages for Texas during the 1980s and 1990s is not representative of the national picture, where the number of active streamgages has remained fairly stable over the last two decades (Figure 2-9) despite some erosion in recent years. The growth in the number of gages through the 1950s and 1960s in Texas was due in part to the extensive surface water development—including reservoir construction—carried out at that time.

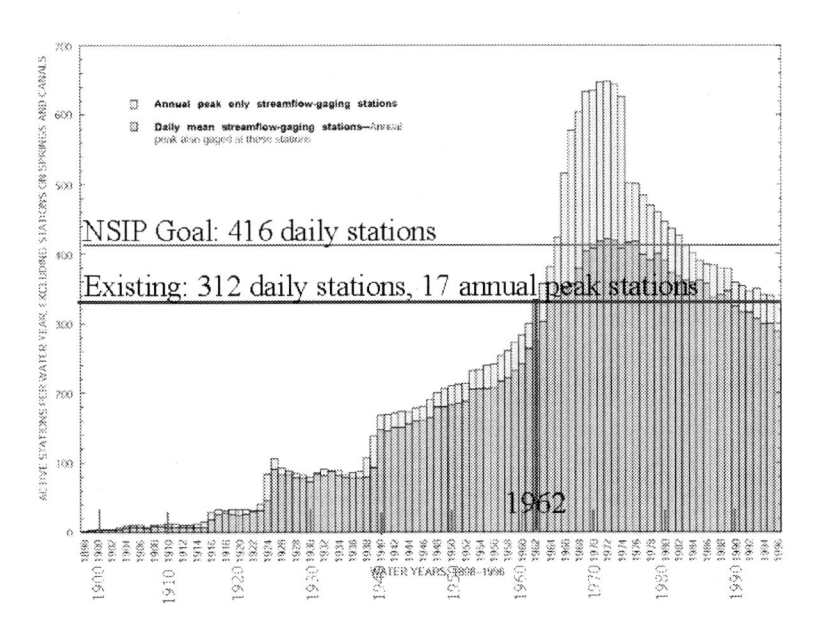

FIGURE 4-1 Trends in streamflow measurement in Texas. SOURCE: Underlying graph from Slade et al. (2001).

Slade et al. (2001) developed a regional optimization model for each of 11 hydrologic regions in Texas (Figure 4-2) using generalized least-squares regression to separate error due to the regression model from error due to a finite sample size. This model estimated mean annual flow and 25-year peak flow using basin characteristics as explanatory variables in multivariate regression equations for each region. Three planning horizons were considered (5 year, 10 year, and 25 year), and active and discontinued stations in natural (i.e., relatively undisturbed) watersheds were considered. In each region, the analysis began with all candidate stations included and then stepped backwards, eliminating the least informative station at each step.

A typical result is shown in Figure 4-3 for estimation of the peak 25-year flow in three hydrologic regions in East Texas. The sampling error was relatively insensitive to the number of stations in the estimation set until this number dropped below about 20 stations, at which point the sampling error started to increase significantly. This figure also shows that as the planning horizon (length of streamgage record) increases from 5 to 10 to 25 years, the sampling error decreases correspondingly. Slade et al. (2001) concluded that

- stations on the steepest part of the curve offered the most valuable regional hydrologic information relative to basin characteristics;
- sampling error increased to the west where the climate is more arid:
  — sampling error for mean annual flow was 6.6 to 114.3%, and
  — sampling error for 25-year peak flow was 9.9 to 28.5%;
- there was greater variability in error between regions than was introdcued by changing the number of stations within a region; and
- there was much less error in regression equations for the 25-year peak flow than for the mean annual flow in arid regions.

Besides the regional regression analysis, Slade et al. (2001) analyzed the correlation among paired stations upstream and downstream of one another on the same river (Figure 4-4). They found the expected strong correlations in flows for upstream and downstream stations on the same river, especially for the mean annual flow:

- 61 of 81 station pairs analyzed for mean annual flow had correlation coefficients > 0.9; and
- 43 of 129 station pairs analyzed for 25-year flow had correlation coefficients > 0.9.

Streamflow Network Design 75

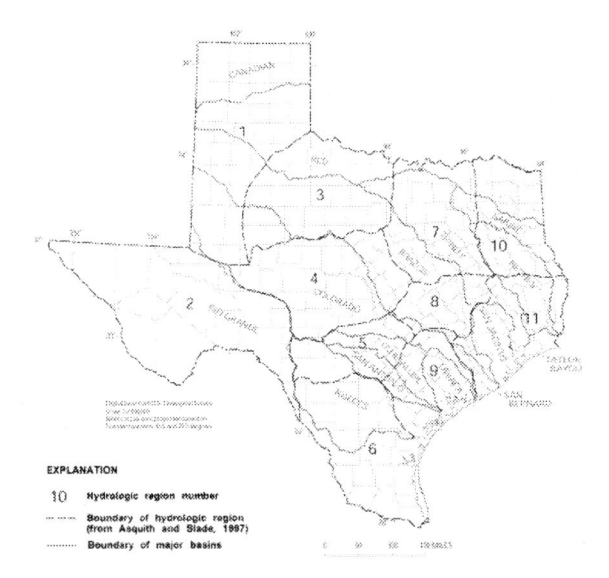

FIGURE 4-2 Hydrologic regions of Texas. SOURCE: Slade et al. (2001).

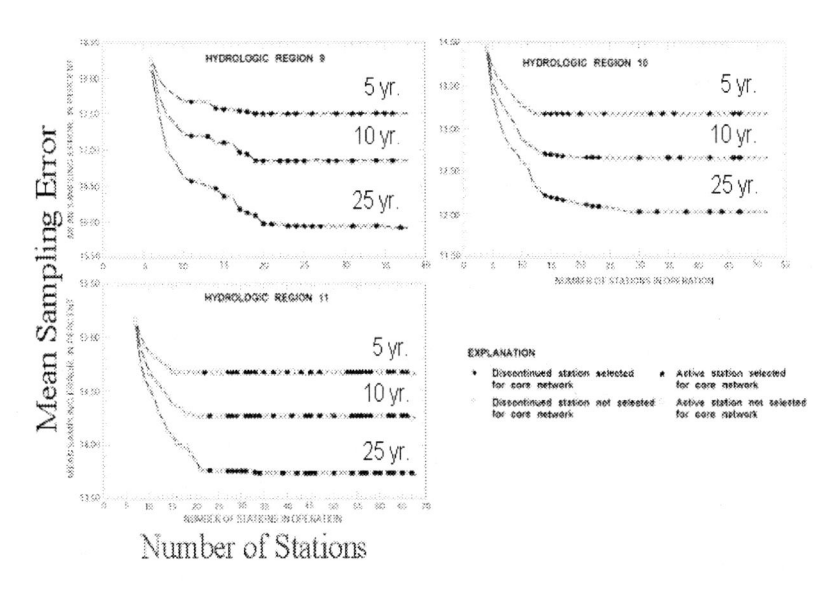

FIGURE 4-3 Sampling error for planning horizons of 5, 10, and 25 years for 25-year peak flow as a function of number of available stations in a hydrologic region. SOURCE: Slade et al. (2001).

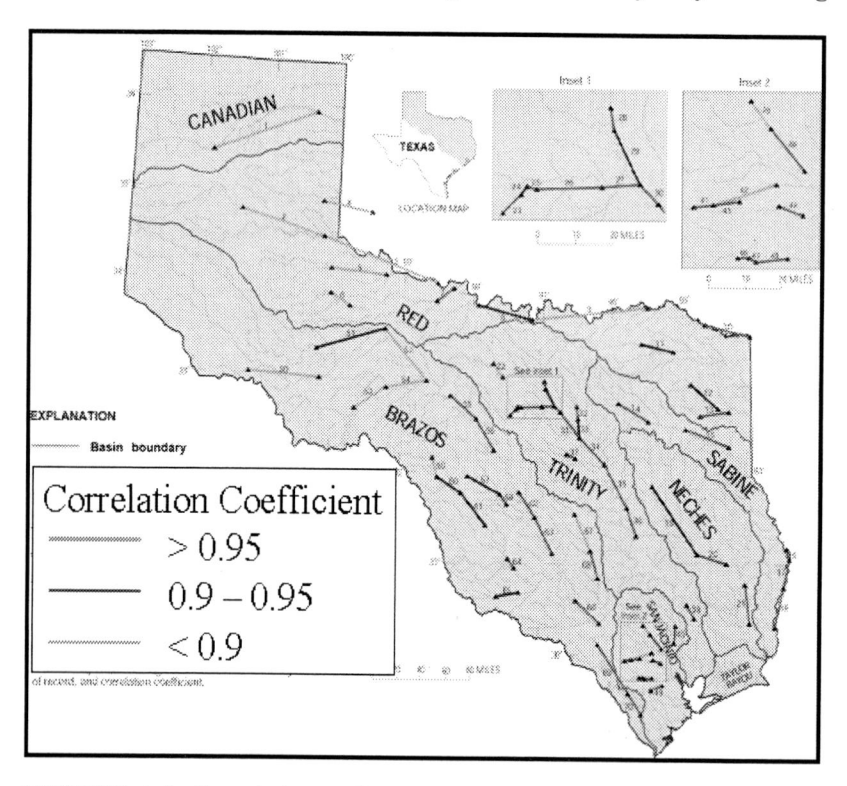

FIGURE 4-4 Correlations of mean annual flow among pairs of stations on the same river. SOURCE: Slade et al. (2001).

As a result, Slade et al. (2001) decided to select stations for a core network that were not highly correlated with other selected stations. The study concluded that Texas needs a core network of 263 stations for regional hydrology purposes on natural watersheds (not including many gages on rivers with large upstream diversions). This number can be contrasted with the NSIP network for Texas, which specifies 416 gage locations. The two numbers, however, are not directly comparable because the statistical study applies to gages in natural watersheds while the NSIP study applies to all watersheds.

This study illustrates both the strengths and the limitations of the statistical approach to network design. The method is rigorous and reproducible, and yields quantitative results about the degree of uncertainty of particular quantiles for a given gage network. The gage sites can thereby be arranged in an unambiguous rank ordering from highest to lowest information content. This helps identify the relative value of each gage for hydrologic regionalization. However, one important limitation of statistical methods is the decoupling of performance metrics used to evaluate network performance from the possibly unrelated purposes for which the gages were installed in the first place. That is, a gage may serve a critical purpose for water management or flood forecasting even if it is not one of the gages most useful for estimating regional hydrologic information at ungaged sites. Although statistical methods can quantify trade-offs between information and cost, such as those in Figure 4-3, these trade-offs (and the value of any particular gage network) change with different design objectives. For example, the "optimal" network to support regional estimation of mean annual discharge (Q_1) and the 25-year discharge (Q_{25}) may differ substantially from the "optimal" network supporting regionalized estimation of 7-day, 10-year low flow ($_7Q_{10}$). More generally, regionalized estimation of a specific set of discharge quantiles ($Q_1, Q_{25}, {}_7Q_{10}$) represents only a small subset of the data generated and information derived from a streamgage network.

Perhaps more significant to the design of a national network, statistical network design methods are most applicable to homogeneous hydrologic regions within which regionalized estimates may be derived. Statistical methods typically assume that the basin response, land use, and climate remain the same over time and may suggest configurations very different from networks designed to detect trends or interventions (Schumacher and Zidek, 1993).

Finally, from a management perspective, statistical methods always yield a "gray" answer rather than a black or white answer as to whether a gage is needed or not. Some gages have more information content, others have less, but it is difficult to know how much information content is enough to justify the existence of a gage.

Statistical methods for stream network design should be used to justify incremental decisions to add or eliminate individual gages within a local gage network serving narrow, well-defined goals (such as hydrologic regionalization). In contrast, the breadth of both the national goals and the hydroclimatic variation spanned by the NSIP network is not meaningfully reduced to a concise set of statistical measures. Thus, the most appropriate role of these methods for the NSIP is supporting the analysis of incremental refinements to local and regional hydrologic networks, within the broader context of NSIP network design.

COVERAGE MODELS

The design of a steamgage network has much in common with a rich family of facility location problems (Drezner, 1995; Drezner and Hamacher, 2002). These include the siting of facilities for fire protection (Schilling et al., 1980; Swersey, 1994), ambulances and hospitals (Branas et al., 2000), vehicle emission test stations (Swersey and Thakur, 1995), hazardous facilities (Kleindorfer and Kunreuther, 1994), oil-spill response centers (Alidi, 1993), and "hubs" (Campbell et al., 2002) for air passengers and cargo transport (Serra et al., 1992).

The concept of a coverage model is best explained by example. Rainfall varies continuously over space, but it can be directly measured only at discrete points (Figure 4-5). Recently, the National Weather Service located a series of weather radar (NEXRAD) sites to estimate this rainfall distribution. Figure 4-6 shows the distribution of NEXRAD radar stations in the 48 conterminous states; each radar provides "coverage" over a range of approximately 200 km, recognizing that the quality of radar coverage degrades with distance. Within an operational definition of "acceptable" coverage, there is a binary aspect to this model in that either an area is covered or it is not. By siting radars so that at least two and preferably three coverages overlap, the National Weather Service (NWS) can observe rainstorms from several angles and estimate the precipitation rate from radar.

Subregions Within Coverage Models

As consequence of defining a coverage model, sampling at discrete locations subdivides a spatial domain into subregions; each subregion is explicitly associated with its respective measurement point. This is typically the case for computing mean areal rainfall from point measurements at raingages, in which Thiessen polygons drawn around the raingage locations are used to estimate watershed average rainfall using an areally weighted average of the raingage values (Figure 4-7). When streamgages are located in a stream network, the watershed draining to that streamgage can analogously be delineated; a unique subarea associated with each gage defines the land area whose drainage flows past that gage before it reaches any other gage (Figure 4-8). This subwatershed is the coverage area associated with that streamgage. Any set of points on a stream network can be used to subdivide a watershed into subwatersheds. Figure 4-9 shows several subwatershed divisions of the Guadalupe basin in Texas for flooding, water quality, and water supply. The upper right panel in this diagram shows

Streamflow Network Design 79

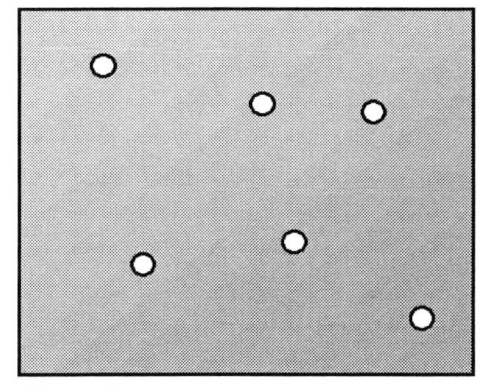

FIGURE 4-5 Coverage of a continuous spatial phenomenon by measurements at points.

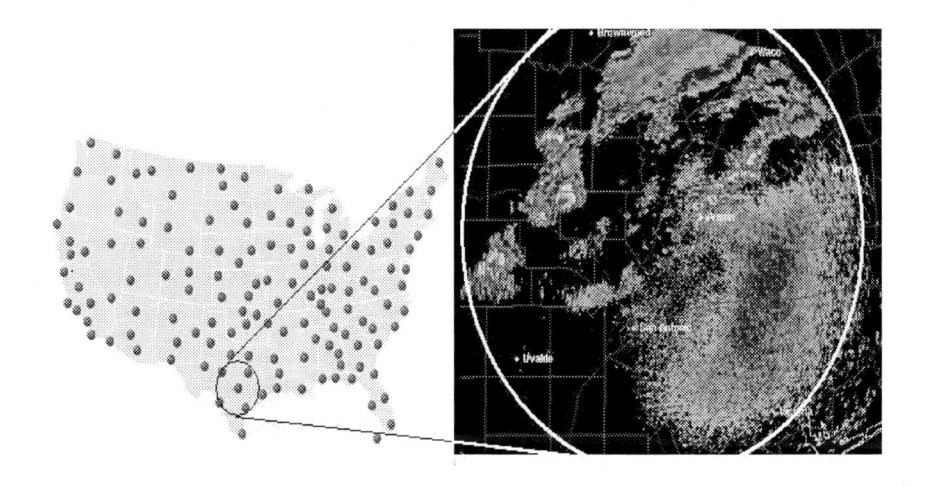

FIGURE 4-6 NEXRAD radar rainfall locations and coverage of radar station KEWX, Austin-San Antonio, Texas. SOURCE: *http://weather.noaa.gov/radar/national.html*.

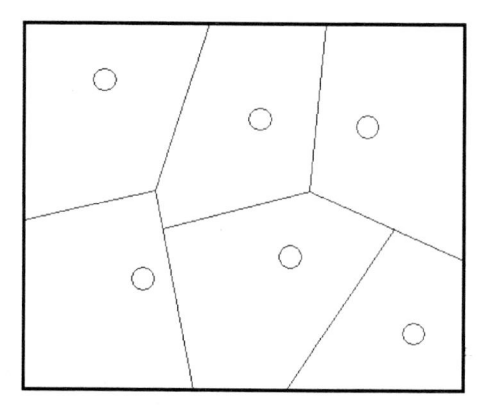

FIGURE 4-7 Spatial subdivision of a region using Thiessen polygons.

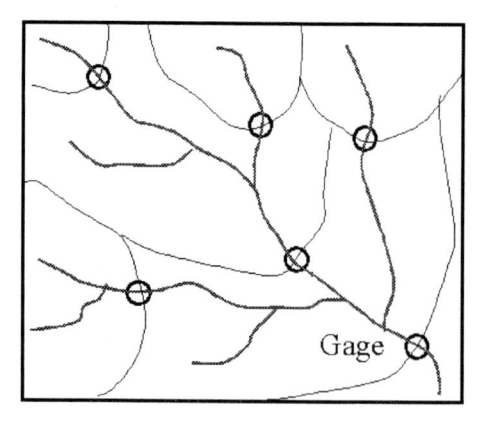

FIGURE 4-8 Spatial subdivision of a region using subwatersheds of streamgages.

the subdivision of the watershed using the NWS river forecast watersheds in which the watershed outlet is an NWS forecast point or data point. The lower right panel shows the subdivision used for the U.S. Environmental Protection Agency's (EPA's) Total Maximum Daily Load (TMDL) studies, where water quality management segments are defined on the principal reaches of the Guadalupe River, and the subwatersheds are the areas draining to these segments. The lower left panel shows the subwatersheds defined for water availability modeling in which the outlet of each subwatershed is a point at which the Texas Commission for Environmental Quality

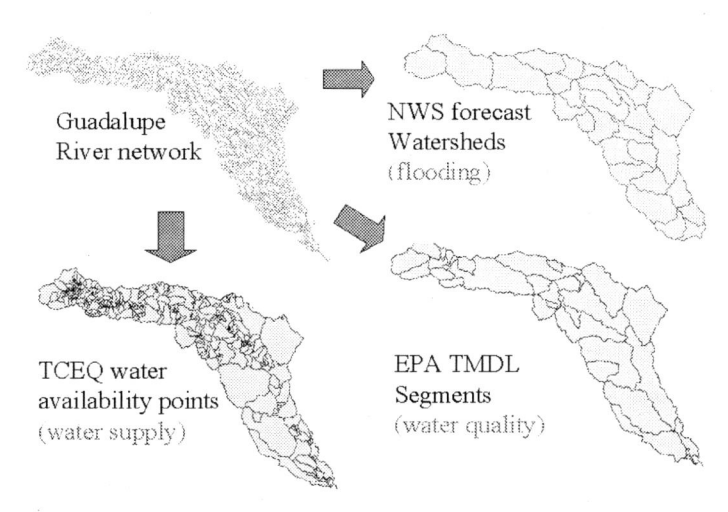

FIGURE 4-9 Subwatershed delineations in the Guadalupe Basin, Texas. SOURCE: Maidment (2002).

(TCEQ) has issued a permit for water withdrawal from the Guadalupe River or its tributaries. As part of estimating the reliability of water supply at these permit points, a long-term water resource simulation is done using monthly data over a period of 40-50 years, in which the "naturalized flow" is estimated for each USGS streamgage (this is the gaged flow adjusted for significant upstream diversions and return flows), and a corresponding naturalized flow is estimated at each diversion point using the ratio of the drainage area of the diversion point and the drainage area of the next downstream streamgage.

In contrast to network designs used to monitor continuous surfaces, fluxes, or fields (e.g., air quality, solar radiation, contaminated groundwater; see Figure 4-5), streamgage locations are confined to the stream network (Figure 4-8), suggesting analogues with facility location in transportation and communication networks. For example, facilities may be optimally sited in a transportation network to intercept traffic flows for vehicle safety inspections or to detect the transportation of hazardous substances (Berman et al., 1995; Gendreau et al., 2000; Hodgson et al., 1996; Mirchandani et al., 1995;). The flow interception location problem engenders subtle trade-offs between maximizing capture (e.g., by locating facilities at the 1995;). The flow interception location problem engenders subtle trade-offs between maximizing capture (e.g., by locating facilities at the "outlet" of directed networks through which all traffic must flow) and "protecting" the network (which favors siting more facilities in the "upstream" reaches of the network for early detection). These problems naturally relate to monitoring and quality management in water distribution networks for which Subramaniam (2001) formulated the location of chlorine booster stations in a water distribution network as a *location set covering problem* (Daskin, 1983).

Service Standards and Thresholds

Many problems with continuous, quantitative performance measures (such as police response time) can be transformed into discrete coverage problems by defining a "service standard." For example, a "threshold" concept of coverage is commonly used to rate residential fire insurance risks, in which a homeowner is considered covered if the home is within 1,000 feet of a fire hydrant or within five miles (or five minutes) of a firehouse. All such homeowners are considered covered and therefore implicitly rated as though they have "equivalent" fire protection, even though homes closer to the fire station clearly have incrementally faster response times. The public interest and public policy in efficiently providing full coverage for critical public services such as fire protection (Marianov and Revelle, 1991) or emergency warning (Current and O'Kelly, 1992) naturally extends to concepts of backup coverage, secondary coverage, and resilience (Haghani, 1996; Hogan and Revelle, 1986; Revelle et al., 1996) in network design.

Where clear accepted service standards can be defined (e.g., insurance standards defining acceptable standards for fire protection) the trade-off between level of coverage and number of facilities (a surrogate for cost) can be meaningfully analyzed. For critical services and national needs, complete, efficient (i.e., minimum number of gages) coverage is the compelling design goal.

An evocative example of the coverage concept to locate a network of facilities was offered by Revelle and Rosing (2000), who analyzed the fourth century deployment of Roman legions by the Emperor Constantine in order to defend (within a particularly defined "level of service") the eight provinces of the Roman empire using only four "field armies." The problem was to either minimize the number of armies required to cover all provinces or maximize the extent of defensive coverage when the number of field armies was inadequate to defend the empire. From the Roman perspective, there was a clear "national" interest in achieving complete coverage of the empire.

THE NSIP NETWORK AS A COVERAGE MODEL

In contrast to the long history of statistically based network design at the USGS, the NSIP network is essentially a coverage model. In its broadest outline, the program has identified a set of gages that satisfies national needs by *covering* "demands" defined by the five NSIP program goals. This long-term design for the national gage network does not attempt to integrate statistical evaluation of the marginal information gains or losses associated with incremental changes in the number and location of gaging stations.

This approach is reasonable. The NSIP network design problem has the complexity of other "strategic network design" (Owen and Daskin, 1998) problems, such as investment decisions to locate international manufacturing facilities that must incorporate future uncertainties and changing conditions. The long-term commitment of limited resources requires such networks to be robust against an uncertain future (Ghosh and McLafferty, 1982; Mulvey et al. 1995; Owen and Daskin, 1998). The design of the national streamgage network must similarly serve current and future national needs and therefore must similarly be designed to be robust against both natural and anthropogenic change. The design for a national gage network is therefore much more complex than the traditional network design problem that has historically been defined by the narrower problem of hydrologic regionalization. Pragmatically, traditional statistical methods based on marginal information value will continue to support incremental decisions and continual improvement in locating new streamgages as the NSIP plan is implemented. **Beyond local refinement, the coverage model based on five minimum national needs is an appropriate model to develop the long-term design of the national streamgage network.**

Some of the NSIP goals, such as gaging for treaty obligations and boundary crossings, are clearly binary coverage goals. For example, the flow of the Colorado River entering Mexico is either gaged or ungaged, and the goal is thereby either covered or not. Other goals, such as gaging river outflows, implicitly define coverage through a service standard (i.e., all basins of a certain size scale; see discussion of goal in Chapter 3).

By analogy to Figure 4-9, the choice of a set of streamgaging sites for each of the five NSIP goals has associated with it a subwatershed dataset that represents the spatial subdivision of the nation into sampling units, each unit having an NSIP gage at its outlet. For three of the NSIP goals (border or compact points, NWS forecast points, water quality points), the point location is chosen first and the subwatershed delineation is determined by these points. For the other two NSIP goals (river basin outflows and sentinel watersheds), the subwatershed dataset is chosen and then streamgaging points are selected at or near the outlets of these subwatersheds. For river basin outflows, the subwatershed dataset is the six-digit USGS hydrologic accounting unit (Figure 3-5), while the sentinel watershed dataset is created by the union of ecoregion boundaries (Figure 3-6) with hydrologic accounting unit boundaries.

Thus, it can be seen that there is a close association between a set of gages chosen to meet an NSIP goal and a subwatershed dataset drawn from these gage points as watershed outlets. The NSIP gage network resulting from the five NSIP goals results in a subwatershed dataset for the nation. In effect, this NSIP subwatershed dataset subdivides the nation into water resources sampling units, each measured by a gage at its outlet.

Since the NSIP base gage site locations for each of the five goals are defined separately for each goal, there does not presently exist a subwatershed dataset that results from all sites taken together. By creating national NSIP subwatershed dataset maps for each criterion using the proposed and active gage sites (approximately 70 percent of the total), the USGS can assess the completeness of coverage. When new gages are to be installed from the NSIP site set, consideration can be given to the impact of site choice on the NSIP subwatershed dataset. The Interstate Council on Water Policy (ICWP, 2002; see following section) suggested that uniformity of coverage, if desirable, could be achieved by locating as many NSIP gages as possible at or near the outlets of the USGS Hydrologic Unit Code (HUC) watersheds, which are part of the Watershed Boundary Dataset of the United States, presently under development. It would also be useful to define the geospatial (e.g., soil and land-use properties, stream network) and hydrologic (e.g., mean annual rainfall and evaporation) properties of these subwatersheds so as to support hydrologic studies of NSIP data with a consistently computed set of supporting watershed data.

The USGS should delineate the subwatershed dataset for the NSIP base gage network stations and define their geospatial and hydrologic properties.

RECOMMENDATIONS OF THE INTERSTATE COUNCIL ON WATER POLICY

The ICWP (2002) assessed the NSIP from the viewpoint of state, local, and tribal users of streamflow data. Because of the importance of the ICWP and its member organizations to state and national streamflow networks, its recommendations are summarized and evaluated.

ICWP Recommendations

In addition to the five NSIP goals, the ICWP considered nine additional goals for streamflow data, originally proposed by the Department of Interior's Advisory Committee on Water Information (ACWI). In doing so, the ICWP not only implicitly accept the validity of the coverage approach taken by the USGS for the program but extended it. These goals include providing (ICWP, 2002, p.1) the following:

- streamflow data for determination of base flood discharges and elevations for communities participating in the National Flood Insurance Program;
- streamflow data for all watersheds with impaired water quality, based on the EPA's TMDL list;
- streamflow data at river reaches with major National Pollution Discharge Elimination System (NPDES) permits;
- stage and discharge information for rivers used for canoeing, kayaking, or rafting;
- streamflow data for rivers draining parcels of federal land of >100 square miles;
- streamflow data for all major rivers with surface water diversions that exceed 25 percent of the river's mean annual flow;
- discharge data for the inflow and outflow of all reservoirs with >50,000 acre-feet of total storage;
- streamflow data for coastal rivers that support a migratory fish population; and
- stage or discharge information on rivers that support commercial navigation.

Like the NSIP network design, a metric was defined for each of these additional nine goals, and the number of gage sites needed to meet these goals was evaluated. The number of sites identified separately for each of the goals totaled more than 30,000, with the largest number of sites supporting National Flood Insurance Program communities (7,297 sites) and Impaired Water Quality Reaches (9,123 sites). Allowing for coincident sites selected by two or more goals, there are 18,330 unique sites chosen according to the 14 goals (the 5 original NSIP goals and the 9 additional goals listed above). It was apparent to the ICWP that not all these goals could be fulfilled by adding new streamgages. Consequently, it recommended the following adjustments to the "base federal network" to be supported by NSIP (ICWP, 2002):

1. Provide stage and discharge data at each National Weather Service and Natural Resource Conservation Service forecast or service location for the purposes of flow forecasting (flood, normal, and drought).
2. Monitor representative discharge at each major subbasin defined as a hydrologic cataloging unit (HUC-8 as opposed to the original HUC-6 basin proposal) for assessing status and trend of flow availability.
3. Provide river streamflow data for rivers governed by compacts between states, tribes, or nations or as dictated under Supreme Court decree (but not including waters crossing jurisdictional boundaries with no legal agreements).
4. Use the existing Hydrologic Benchmark (HBM) station network to monitor streamflow and act as sentinel watersheds to evaluate altered rainfall-runoff relations induced by changes in climate or weather.

The ICWP also recommended what it called "a new concept: defining a national network through watershed coverage." This would involve subdividing the landscape into HUC-8 and HUC-10/11 subwatersheds and siting gages funded by the Cooperative Water Program at or near the terminus of each HUC-8 subbasin and, within these subbasins, have gages placed as a function of the localized water management need for such information. For example, Kansas has 12 HUC-6 units, 80 HUC-8 subbasins, and 330 HUC-10/11 units, and presently has 166 NSIP gage locations identified. In the coverage model proposed by the ICWP, federal-state cooperative gages would be sited in such a manner as to augment the NSIP distribution and be representative of all HUC-8 and as many HUC-10/11 units as possible.

The ICWP concept of identifying gage locations by a coverage subwatershed model is consistent with the design of the national gage network proposed by the NSIP. Using subwatershed coverage to locate streamgages is an appropriate approach to designing a robust national network and is similarly endorsed by the committee.

Comments on the ICWP Recommendations

Providing additional feedback on the ICWP recommendations requires that one first make an important distinction between *data collection* (or, specifically, *streamgaging) points* and *information points*. The former are locations at which streamflow and(or) some other property is measured; the latter represent sites at which streamflow information is desired and generated from the available data. Advances in geospatial information technology in conjunction with the National Hydrography Dataset, the National Elevation Dataset, and modeling techniques have greatly improved our accuracy in spatially estimating streamflow (with confidence limits) for a dataset of information points on the stream network. Applications of this concept are further developed later in this chapter.

ICWP recommendation 1: "Provide stage and discharge data at each National Weather Service and Natural Resource Conservation Service forecast or service location for the purposes of flow forecasting (flood, normal and drought)."

The committee concurs with this recommendation to include the Natural Resource Conservation Service (i.e., not just NWS) forecast points as part of the NSIP flow forecasting goal information points.

ICWP recommendation 2: "Monitor representative discharge at each major sub-basin defined as a hydrologic cataloging unit (HUC-8 as opposed to the original HUC-6 basin proposal) for assessing status and trend of flow availability."

The six-digit HUC is an appropriate scale to characterize flows of the nation's major rivers and evaluate national river outflows from the continental United States. There are many uses and a clear national need for streamflow information from the smaller, eight-digit and ten-digit HUCs as well. Encouraging cooperators to support gages at the outlets of HUC-8 and HUC-10 watersheds is a desirable goal. Pragmatically it is unclear that the national needs for streamflow information from eight- and ten-digit HUCs can be reliably satisfied opportunistically, within the Cooperative Water Program. The USGS should therefore consider a stratified random sampling design to gage and characterize smaller watersheds. This design should support and be closely coordinated with methods development to provide consistent estimates of streamflow information for all eight- and ten-digit HUCs.

The provision of streamflow information at boundaries of standardized watersheds is desirable, and the HUC-8 dataset, and the emerging HUC-10 and HUC-12 datasets from the Watershed Boundary Dataset, should be considered *information points* if not specifically gaging sites. The USGS should develop a coverage-based method to provide streamflow information *with quantitative confidence limits* for these information points using an appropriate combination of measurement technologies, data assimilation, and synthesis techniques.

ICWP recommendation 3: "Provide river streamflow data for rivers governed by compacts between states, tribes or nations or as dictated under Supreme Court decree (but not including waters crossing jurisdictional boundaries with no legal agreements)."

In its report on the USGS National Water-Use Information Program (NRC, 2002), this committee documented the status of legal permitting for water use in all 50 states. Rules and legal procedures differ significantly from state to state, and conflicts have arisen among several states over shared waters crossing state boundaries. As the intensity of water use increases in the future, more conflicts of this kind may be expected. In that event, long-term streamflow records from the USGS as an independent, trusted source of information will be required. Further, even if no legal conflict between states develops, state water availability planning requires the capacity to separate water arising from within the state from that flowing into the state.

The committee does not concur with the ICWP recommendation to eliminate from the NSIP program gage sites on jurisdictional boundaries with no legal agreements.

ICWP Recommendation 4: "Use the existing Hydrologic Benchmark station network to monitor streamflow and act as sentinel watersheds to evaluate altered rainfall-runoff relations induced by changes in climate or weather."

The HBM network is a set of 73 gage locations in pristine environments intended to monitor flows in undisturbed watersheds. The sentinel watershed goal of the NSIP generates 874 gage site locations representative of the nation's ecological and hydrologic regimes. This broad distribution of representative sites is valuable and represents much more than relatively pristine catchments with minimal human influence. Although sentinel watershed gages are chosen to be relatively unaffected by flow regulation and diversions, they are specifically selected to characterize the ever-changing status of the nation's water resources in response to changes in climate, land use, and water use in 800 watersheds that typify major ecoregions and river basins.

The current sentinel watershed goal sites should be retained rather than just using the Hydrologic Benchmark sites.

Concerning the additional nine goals identified by the ACWI and examined by the ICWP, and the total of 18,330 gage sites thus located, all of the ACWI-identified goals have merit. In particular, the goals of supporting

Impaired Water Quality Reaches for TMDL studies (9,123 sites) and National Flood Insurance Program communities (7,297 sites) have national significance, directly supporting federal water quality and flood mitigation programs. Gaging all sites required to meet these goals would be well beyond the capacity of a national network, even under the most optimistic assumptions about future funding. However, having a streamgage at each information point is not the only way to provide streamflow information. Further, streamgages are but one of many different data collection technologies that can be used to support the generation of streamflow information.

The additional sites identified to serve ICWP goals represent significant valuable information needs and should be considered information points. The USGS should develop a coverage-based method to provide streamflow information *with quantitative confidence limits* for these information points using an appropriate combination of measurement technologies, data assimilation, and synthesis techniques.

NETWORK DESIGN GOALS: CONTRASTING NSIP WITH STATE-DESIGNED STREAMFLOW NETWORKS

During the 1980s the USGS sponsored several state-level studies assessing the adequacy of the state's streamgage networks (e.g., Fontaine et al., 1984; Medina, 1987). The prototype for these studies was the USGS network in Maine, where "the stream gaging activity is no longer considered a network of observation points, but rather an information system in which data are provided by both observation and synthesis" (Fontaine et al., 1984). A typical set of goals from these studies is listed below:

- *Regional hydrology*—relating basin characteristics to streamflow under natural conditions
- *Hydrologic systems*—water accounting including diversions and return flows
- *Legal obligations*—treaties, compacts, and decrees
- *Planning and design*—dams, levees, and water supply
- *Project operation*—reservoir releases and hydropower
- *Hydrologic forecasts*—floods and flow volumes

- *Water quality monitoring*—National Stream Quality Accounting Network
- *Research*—gages for specific studies
- *Other*—recreation (e.g., canoeists, fishermen)

This list of goals is more extensive than the goals adopted by the NSIP, but a side-by-side comparison of the two lists in Table 4-1 indicates that they have a good deal of commonality. The goals from the above list that are omitted in the NSIP are planning and design of facilities, project operation, research, and other purposes such as recreation and canoeing.

In considering goals such as the operation of facilities or research on a particular watershed, a disproportionate share of information value may go to a limited set of well-identified stakeholders. Similarly, recreational users of streamflow information are important locally, but streamgages designed to serve these needs may be difficult to justify at the national level. These disparities make local partners strong candidates for cooperative funding and other innovative arrangements to support the gage network. Consequently, the USGS has responded to uncertainty and variability in streamgage funding with vigorous and creative development of cooperative funding arrangements to avoid eliminating gaging stations. Indeed, one of the concerns that prompted establishment of the NSIP was the unreliability of funding from agencies operating water facilities. Nevertheless, in considering *national* needs supported by the NSIP network, valuable local and regional goals such as specific watershed research or operational needs should not play an overriding role in national network design.

Another area that is often mentioned as a candidate for the NSIP is urban hydrology. Land-use change associated with population growth is a broadly national issue, and this committee endorses Goal 4 of the NSIP (using sentinel watersheds to regionalize streamflow characteristics and assess trends in streamflow due to factors such as changes in climate, land use, and water use); see Chapter 3. However, many of the more specific goals for streamflow measurement in urban areas are not appropriate for a national program or are not appropriate for a USGS program.

For example, measurement of flow and water quality from large sewer pipes whose discharge is regulated by EPA may be most appropriately performed by that agency or a state regulatory agency. Short-term measurement of flows at street and highway crossings to generate design data for culverts might be done more appropriately by federal, state, or local highway administrations. Also, regulatory authority over stormwater and erosion issues is a local, not a federal, matter. Thus, streamgaging in urban areas is often driven by regulatory reasons, by transportation interests, or

TABLE 4-1 Comparison of NSIP Network Design Goals with Those of Earlier State Network Design Studies

1980s Network Design Goals	NSIP Network Design Goals
Regional hydrology	Sentinel watersheds
Hydrologic systems	River basin outflows
Legal obligations	Borders and compacts
Planning and design	No
Project operation	No
Hydrologic forecasts	NWS flow forecasts
Water quality monitoring	Water quality
Research	No
Other (recreation, canoeing)	No

simply by the desire of a city administration to manage its streams and watersheds. It is unclear that there is a major federal interest in many of these activities and, where there is a federal interest, that the USGS is the best agency to assume the responsibility.

Another area of streamflow measurement that is of concern, especially from the viewpoint of river science, is to gage very small, first- or second-order headwater streams. These small streams are critical components of river networks. Although gaging such small streams is part of the USGS research program, as at the Luquillo Experimental Forest in Puerto Rico (http://pr.water.usgs.gov/public/webb/), it is not undertaken generally around the nation and gaging small watersheds is not an explicit part of the NSIP. If a GIS (geographic information system) based metric for gaging small streams were to be developed similar to the other five goals in NSIP, it would require a high-resolution digital representation of the stream network of the nation. At present, the best representation of the digital stream network of the nation is the National Hydrography Dataset (NHD) at the relatively coarse scale of 1:100,000. For some states, 1:24,000-scale NHD data have been or are being prepared. The definition of what constitutes first- and second-order streams changes with the scale of the map representation, with the higher-resolution 1:24,000 data yielding a larger number of smaller first-order streams than the 1:100,000-scale data.

Thus, the digital basis for systematically defining first- and second-order streams across the nation is improving but is not yet in place. However, the USGS should revisit the issue of gaging first- and second-order streams in the future as part of its review process, as the degree of detail of the geospatial coverage of the nation's streams continues to improve. This might be done through random subsampling of small watersheds with the cooperation of other agencies and the private sector (e.g., transportation).

In addition to Maine, other statewide analyses have been done in recent years and have taken a variety of approaches. These include studies of the Wisconsin (Team for Evaluating the Wisconsin Water-Monitoring Network, 1998), Delaware (Doheny, 1998), Maryland (Cleaves and Doheny, 2000), Illinois (Knapp and Markus, 2003), and Texas (Slade et al., 2001) networks. The Wisconsin study uses Geographic Management Units established by the Wisconsin Department of Natural Resources as its basic watershed coverage for streamgage planning. The Delaware study cites a list of goals similar to those given above for streamgaging in Maine. The Maryland program attempts to cover various water management goals while maintaining long-term gages and a broad range of geographic areas and watershed sizes. Illinois has focused on understanding the many needs of users with an exhaustive survey of both the public and the private sector. It acknowledges the impossibility of anticipating many of the future data needs of the program and therefore supports maintaining a base network that is "representative of the streams of Illinois, such that these long-term data are available to meet a broad range of potential needs" (Knapp and Markus, 2003). The Texas study is discussed in detail earlier in this chapter as an example of statistical network design.

NSIP NETWORK DESIGN: FROM DATA TO INFORMATION

As noted earlier in this chapter, there is a sharp distinction between sets of gaging points (i.e., sites at which streamflow is measured) and sets of information points (i.e., sites at which streamflow information is generated). These sets are not mutually exclusive. This distinction mirrors Fontaine et al.'s (1984) description of the Maine streamgage program as an information program supported by both observation and synthesis. As a national *information* program, the NSIP is the primary federal program to satisfy the nation's current and future needs for streamflow information, supported by both observation and synthesis. The broad long-term goals of the NSIP should be building the capacity to provide streamflow information (with rigorous, quantitative confidence limits) at any arbitrary information point in the nation.

The NSIP should be integrated, managed, and evaluated as a national information program, strategically focused on the long-term goal of providing streamflow information with confidence limits at any arbitrary point in the landscape. The design and continuous refinement of the NSIP gage network should be driven by and consistent with this broad overarching goal.

Quality and Value of Information

Emphasizing both information and confidence limits acknowledges that streamflow information is generated through a suite of measurement technologies and synthesis methods that jointly determine the quality of information. Here information *quality* and information *value* must be distinguished; the value of information can be determined only in the context of applications and decision making supported by that information (Cleveland and Yeh, 1990; Wagner, 1999). For example, real-time streamflow information can be a critical component in flash flood warning and response, yet the marginal value of gage information cannot be quantified independently from the warning, dissemination, and emergency response plans that collectively determine the effectiveness of any flash flood warning system (Drabek, 1999; Gruntfest and Handmer, 2001; Handmer et al., 1999).

For this reason the value of streamflow information is inherently coupled to its many and growing uses, as national demands for streamflow information change. The evolving needs for streamflow information are illustrated in the prioritization of FY 2003 streamflow information needs within the Cooperative Water (Coop) Program (*http://water.usgs.gov/coop/priorities.-html*). The general category of hydrologic hazards has been a core focus of the USGS for many years. However, the recent dramatic fires in the western United States have highlighted the need for improved understanding of the effects of "large-scale forest fires," which is now explicitly identified among the Coop priorities. While suggesting the potential capacity for the Coop program to respond to emerging needs, if funding is available, this also highlights the need for a robust capacity for adaptation within the NSIP data collection program.

If cooperative funding is available, opportunistic data collection directed to watersheds experiencing large-scale forest fires will provide a wealth of information ranging from understanding sediment storage, disturbance ecology, and the biogeochemical cycles in fire-disturbed ecosystems, to practical management information on changes in flood risks and sedimentation. However, the value of these data would be greatly increased if baseline data collection had been initiated prior to these extreme events. Data collection to establish baseline conditions in anticipation of future, uncertain needs may be particularly difficult to support through the Cooperative Water Program. Of course any decision to collect such baseline data must anticipate its future use. Strategic anticipation of future needs is more appropriately incorporated into core federally funded NSIP data collection efforts.

Consider for example, the Hydroclimatic Data Network (HCDN) consisting of USGS streamgages with relatively long records on watersheds that are minimally affected by regulation and diversions (Slack and Landwehr, 1992). This unique network has proven especially useful in evaluating hydrologic trends and testing climate change hypotheses (Lins, 1997; Lins and Slack, 1999; McCabe and Wolock, 2002; Vogel et al., 1999). Though highly valued today, the HCDN is a "discovered" network that exists today only as the cumulative result of a series of independent gaging decisions made over the last century. When decisions were made to support these gage stations, their future use in the analysis of climate change was unimagined. Moreover, although the marginal value of adding additional gages with long records in "natural" watersheds could be estimated, it is too late to add these gages today—regardless of their value.

The current discovered value of the HCDN gages illustrates the importance of considering the nation's future needs and future uses for streamflow information. The challenging decision to commit current gaging resources that will support the nation's future (and uncertain) needs for streamflow information does not lend itself to traditional cost-benefit analysis. Predicting future streamflow needs with certainty is obviously not possible. Although the particular needs that will emerge in the future cannot be confidently predicted, one can confidently predict that such needs will emerge.

Thus, as first noted in Chapter 3, the NSIP program should therefore be structured with the robust capacity to target data collection resources to likely future needs. For example, powerful trends in population growth (and accompanying water use) in the arid Southwest and near the coastal ocean portend future demands and the likely value of "current" baseline water information in these hydroclimatic regions. The NSIP should also support data collection for less certain future needs for expanded data collection, such as enhanced streamflow information in coastal zone streams discharging to estuaries or ephemeral streams in the Great Basin.

The USGS should create a mechanism to institutionalize adaptive management of the nation's likely future needs for streamflow information and provide a mechanism to support these likely emerging needs as part of the core federally funded gage network.

NSIP: An Enhanced National Information System

When viewed as an information program supported by observation and synthesis, the NSIP motivates a new paradigm for streamflow data collection and management. The current model emphasizes data collection and processing of stage height measurements that are synthesized, electronically archived, and most commonly reported as discharge values. Storage and dissemination of streamflow information are primarily oriented to tabular values of daily average discharge reported at the location of a streamgage. In contrast, the NSIP should support an "information base" that is both spatially and substantively far more expansive. Spatially, the goal of providing streamflow information at an arbitrary point in the landscape generalizes the concept of information points and requires close integration of data collection, data management, and methods development for information generation. Substantively, the national need for streamflow information extends far beyond discharge measurements and includes information about the geomorphic characteristics of the stream channel, the riparian corridor, the landscape, and their coupled biogeochemical and ecological systems. While maintaining continuity with historical and current gaging technologies, the application of the nation's streamflow information program to evolving societal needs such as river science (see Chapter 6) will demand new paradigms in data collection and data management, as well as a consistently rigorous approach to the generation, management, and dissemination of information.

Conceived in this way, the dynamic NSIP can be viewed as supporting a continuous streamflow "information cycle," represented conceptually in Figure 4-10. Built on the USGS's core expertise in streamflow measurement, NSIP *data collection* relies on the NSIP gage network, including a base network of federally funded gaging stations. However, data collection also integrates the full range of data collection technologies and procedures, including crest stage gages, intensive data collection during hydrologic extremes, remote sensing, and innovative technologies for non-contact water measurement. Moreover, NSIP data collection involves far more than discharge measurements and includes a broader suite of measurements within the channel (e.g., velocity fields, bed material load, channel geometry, stream biota) as well as measurements that characterize the form and function of the riparian corridor and floodplain.

This richer data collection stream requires a *data management* system with the capacity to handle very diverse data formats, ranging from remotely sensed digital imagery to four-dimensional velocity fields derived from acoustic Doppler current measurements over a river reach. Together, the

Streamflow Information Cycle

Data Collection
- Stream gages
- Intense data collection during extreme events
- Water quality

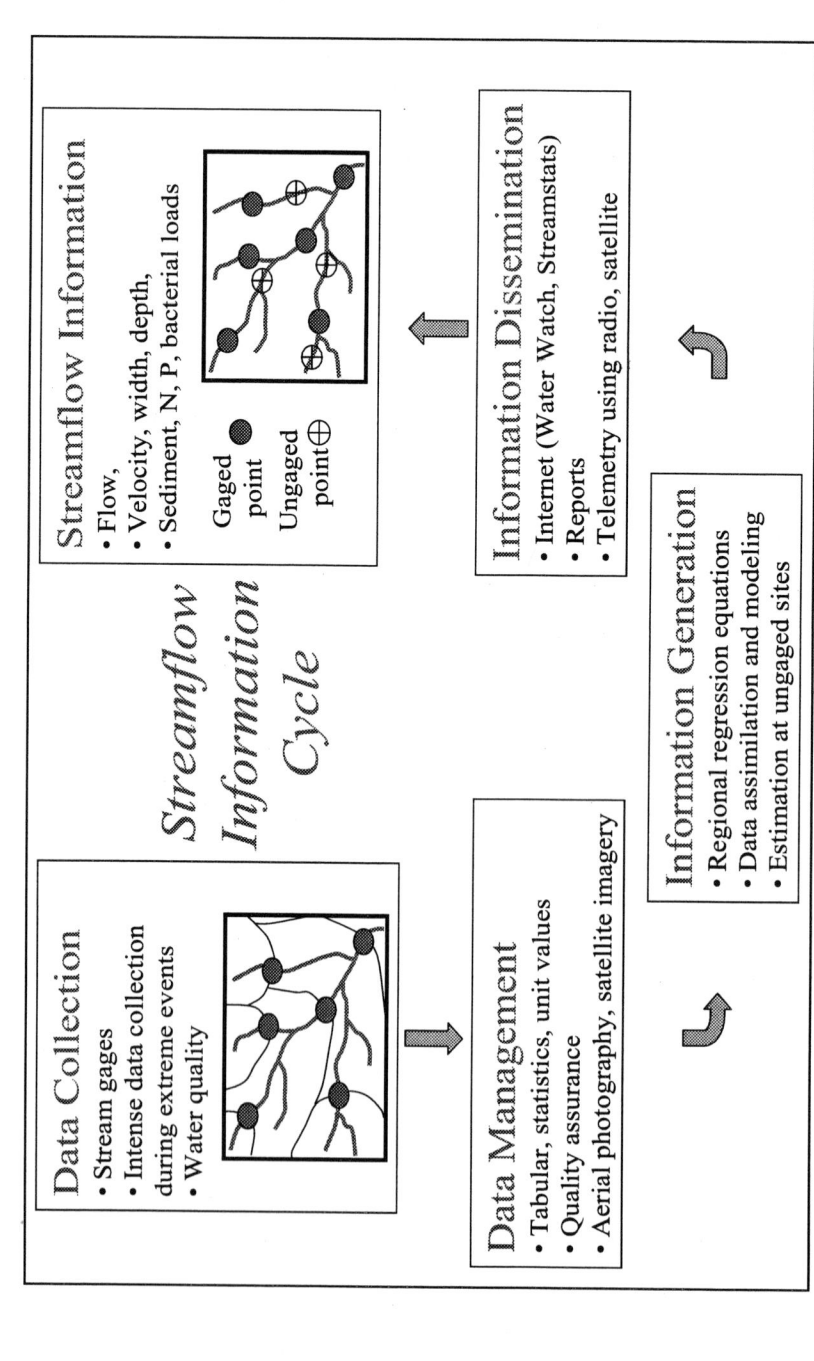

Data Management
- Tabular, statistics, unit values
- Quality assurance
- Aerial photography, satellite imagery

Information Generation
- Regional regression equations
- Data assimilation and modeling
- Estimation at ungaged sites

Information Dissemination
- Internet (Water Watch, Streamstats)
- Reports
- Telemetry using radio, satellite

Streamflow Information
- Flow,
- Velocity, width, depth,
- Sediment, N, P, bacterial loads

Gaged point ● Ungaged point ⊕

FIGURE 4-10 Streamflow information cycle: from data to information.

data collection and data management components of the NSIP support *information generation*, providing streamflow information, with quantitative confidence limits, at any information point in the landscape. NSIP information generation incorporates traditional hydrologic regionalization and statistical approaches for estimating discharge at ungaged sites, as well as methods development to incorporate spatially referenced information (e.g., land use, land cover, topography, water control structures) and indirect information such as paleoflood deposits and historical high-water marks.

Like the expanded scope of the data collection and data management components of the NSIP, *information dissemination* should expand the USGS's exceptional commitment to the Internet and extend to other emerging information technologies and models for information dissemination. For example, the current USGS technology for information dissemination is a user "pull" moder, in which users can access, select, and download streamflow information. Alternate models allow users to specify data needs that may be accumulated passively from a larger data stream using "push" technologies, that is, data is transferred as the data stream is generated without requiring user action. Push technologies have been successfully developed and economically deployed using satellite, radio, and the Internet by, for example, the NWS to support the Emergency Managers Weather Information Network.

Together these NSIP components provide the framework to support the nation's expanding need for *streamflow information*. The streamflow information cycle is then "closed" by continuous feedback and the recurring systematic evaluation of current and emerging information needs. Generating streamflow information with quantitative confidence limits helps both in its interpretation as well as in linking the quality of the information to its value for individual users and the nation.

It is recognized that in the past, watershed information has been neither the traditional nor the primary goal of the USGS streamgaging program. However, the NSIP will establish the observational and data infrastructure for the nation's streamflow information needs in years to come. The USGS should therefore anticipate the needs for streamflow information to address emerging science questions ranging from the source, flowpaths, and dominant mechanisms of overland flow to the role of hyporheic processes in the fate and transport of nutrients and contaminants. As the nation's streamflow information program, the NSIP can anticipate and lay the foundation for the continued development of integrated "river science" programs within the USGS and at other institutions (see Chapter 6).

SUMMARY

The USGS has been exceptionally successful as the nation's source for unbiased, science-based water resources information, despite great uncertainty and variability in funding for basic, core data collection and continuous operation of the national streamgage network. The USGS's responsibility to meet current and future national needs requires a strategic network design (Owen and Daskin, 1998) structured to be robust against inevitable changes and uncertainty. The network should be oriented toward the overarching goal of providing streamflow information with confidence limits at an arbitrary information point in the landscape. Tactically, both limited funding and changing needs will require the USGS to continually reevaluate, refine, and adjust the national gage network. Success can only be judged iteratively and will require continual refinement of the network.

Many approaches have been used to design and maintain data collection networks. Statistical procedures offer numerical precision for network design and quantitative estimates of uncertainty. However, they are most effective in local to regional, homogeneous regions, and they do not support the many other goals and uses of site-specific streamflow data. In contrast, coverage models that articulate a goal, define a metric that identifies locations supporting that goal, and apply this procedure to yield a set of potential sites for gages, have many advantages for a national network.

Each of the NSIP components contributes to both the quality and the value of streamflow information. This streamflow information cycle should, of course, represent an ongoing process of evaluation and improvement. Overall, the proposed design of the NSIP streamgage network represents a sound and well-reasoned foundation to support this continuous process.

The use of a coverage model to design the national gage network to meet the five NSIP goals represents a sound approach to designing a robust data collection network for the NSIP. Where possible, statistical methods that quantify the marginal information gains or losses from incremental changes in local and regional gage networks should be integrated into the implementation of the NSIP plan, including the continual refinement and reevaluation of hydrologic data collection. The NSIP program should include an explicit mechanism to direct gaging resources to support emerging issues of national significance.

The NSIP's current model emphasizes data collection and processing of stage measurements that are synthesized, electronically archived, and most commonly reported as discharge values. However, the NSIP should support an "information base" that is both spatially and substantively far more expansive. Its goal should be providing streamflow information at any arbitrary point in the landscape, and this information should include information about the geomorphic characteristics of the stream channel, the riparian corridor, the landscape, and their coupled biogeochemical and ecological systems whenever feasible. **The program should support a continuous streamflow "information cycle" of data collection, data management, information generation, and information dissemination.**

This richer data collection stream requires a data management system with the capacity to handle very diverse data formats, ranging from remotely sensed digital imagery to four-dimensional velocity fields.

Generating streamflow information with quantitative confidence limits is important in linking the quality of NSIP streamflow information to its value both individual users and the nation.

5
Streamflow Information

The National Streamflow Information Program (NSIP) is more than a streamgaging program. It is a comprehensive program designed to provide high quality and accessible streamflow information suitable for multiple uses (USGS, 1999). In addition to the nationwide system of federal interest streamgaging stations for measuring streamflow and related environmental variables, the NSIP has four other components:

1. a program for intensive data collection in response to major floods and droughts;
2. a program for periodic assessments and interpretation of streamflow data to better define their statistical characteristics and trends;
3. a system for real-time streamflow information delivery to customers that includes data processing, quality assurance, archiving, and access; and
4. a program of techniques development and research.

The purpose of this chapter is to summarize and assess the activities that the U.S. Geological Survey (USGS) has initiated to address these components. It should be noted that the full scope of the various subject areas covered in this chapter is extensive. The purpose of the chapter is not to survey all work done in these fields, but rather to summarize of the various studies and techniques that were presented by the USGS to the committee during the course of its study and to comment on the value of these activities.

INTENSE DATA COLLECTION DURING FLOODS AND DROUGHTS

As described in USGS (1999), "The NSIP approach to data collection for floods and droughts will be to supplement data from streamgaging stations with systematic field surveys. Every flood and drought is unique, but a standardized approach to field work and data collection will ensure that the important aspects of each event are documented. Data collected during these events will include information about precipitation duration/frequency, river stage and discharge, and opportunistic sampling of water quality variables to include suspended sediment, nutrients, specific conductance, alkalinity, bacteria, pesticides, and hydrocarbons. Changes in the geomorphology of river channels, such as river-bank erosion location and processes, and sedimentation volumes and distribution would be documented for high- as well as low-flow conditions."

Intense Monitoring During Floods

Streamflow conditions during floods are materially different from those during normal or low flows because the stream is no longer confined within its channel and may range widely over the floodplain (Figure 5-1). It is during floods that most of the annual sediment load is transported, and because many contaminants are adhered to sediments, floods are also a significant transporter of contaminants.

A possible prototype for the study and documentation of a major hydrologic event is demonstrated in U.S. Geological Survey Circular 1120, *Floods in the Upper Mississippi River Basin* (available on-line at *http://water.usgs.gov/pubs/circ/*). This circular series, with 12 chapters published between 1993 and 1998, provided a timely synopsis and assessment of the effects of the 1993 Midwest floods. After a wet spring, widespread flooding was caused by a persistent anomalous weather pattern in the summer, which produced excessive rainfall throughout a nine-state area (Wahl et al., 1995). Unusual aspects of the flood event that were identified included the large region affected by record flooding, especially during the summer season, and the long duration of the floods (Parrett et al., 1993). Relying heavily on data gathered at USGS streamgages, as well as special data collection efforts during and after the flooding, USGS Circular 1120 documented the magnitude and frequency of peak discharges and flood volumes (Eash, 1997; Moody, 1995; Parrett et al., 1993; Southard, 1995); the effects of reservoir storage on flood peaks (Perry, 1994); water quality characteristics of floods,

FIGURE 5-1 The Willamette River Flood, 1996. SOURCE: Bonneville Power Administration (http://www.bpa.gov/Power/pl/columbia/4-gal-2.htm).

such as chemical and sediment transport and deposition (Goolsby et al., 1993; Holmes, 1996; Schalk et al., 1998; Taylor et al., 1994); and the effects of inundation on groundwater quality (Kolpin and Thurman, 1995), as well as geomorphologic changes and stream-channel scour at bridges (Jacobson and Oberg, 1997). The series is noteworthy for more than its content; its publication so soon after the floods (the first five chapters were published within six months of the event) significantly enhanced its impact on the public and the scientific community.

An important contribution of documenting the 1993 Upper Mississippi floods was its impact on the scientific study of flood processes. In particular, some of the findings challenge conventional wisdom on the role of major floods in the transport of agricultural chemicals from the landscape (Goolsby et al., 1993). Although runoff during floods transports large amounts of nutrients, herbicides, and other agricultural chemicals to rivers, flood waters are thought to dilute the chemicals, resulting in lower chemical concentrations. However, a comparison of measurements showed that herbicide concentrations during the spring and summer of 1993 were similar to the maximum concentrations observed in the spring and summer of 1991 and 1992. Furthermore, water quality measurements showed that total chemical loads to the Gulf of Mexico during the spring and summer of 1993 were significantly larger than those in 1991 (80 percent larger for atrazine and 37 percent for nitrate nitrogen) and up to several times larger than in 1992 (235 percent larger for atrazine and 112 percent larger nitrate-nitrogen). Goolsby et al. (1993) concluded that the high loads of nitrates into the Gulf of Mexico could increase phytoplankton biomass, affecting the ecosystem along the Louisiana coast.

Data collection during major floods is challenging. Ironically, it sometimes happens that streamgaging stations are washed out during peak high-flow events when their records are most needed. Furthermore, the nature of floods means that direct access to streams for measurements may be difficult or hazardous. Remote sensing may offer innovative ways of gathering information on the extent of inundation over large areas or sediment concentration and loads during major floods. For example, it is possible to use satellite remote sensing on clear days to record the extent of inundation during regional flooding, and also to use radar measurement from aircraft during both night and day to sense the extent of surface water inundation. Since radar penetrates, clouds it is feasible to operate with this technique in adverse weather conditions. It may even be possible to routinely monitor regional floods from unmanned aerial vehicles similar to the drones employed during military campaigns.

Since the extent and depth of flood inundation are the critical factors causing flood damage, remotely sensed images of flood inundation from space, coupled with an accurate terrain surface model, would allow computation of the volume of water inundation. If a regularly sequenced set of such images were obtained, and corresponding volumes calculated, data for verifying two-dimensional models of flood inundation could be obtained, and perhaps new types of flood propagation models could be developed using finite volume methods. For example, Alsdorf et al. (2000) and Alsdorf (2003) used interferometric radar measurements of water-level changes on the Amazon floodplain to calculate volume changes, from which average discharge rates could be deduced. Smith and Alsdorf (1997) similarly used decorrelation of tandem European Remote Sensing Satellite (ERS) data to map flooding changes on the Ob River in Siberia, and Mertes et al. (1993) used Landsat images to estimate suspended sediment concentrations in the Amazon River.

Intense Monitoring During Droughts

Droughts offer the opportunity to quantify the low-flow characteristics of streams and rivers. This is typically done by establishing a network of secondary and tertiary streamgaging sites and conducting regular streamgaging surveys of them (Hardison and Moss, 1972; Riggs, 1972). A secondary site is one where a gage board has been installed and periodic measurement of stage is undertaken but no continuous recorder is installed. A tertiary site is one where no stage record is maintained, but rather the site is used solely for periodic streamflow measurement by current meters, acoustic Doppler current profilers, or perhaps in the future by noncontact land-based remote sensing approaches (see "Methods Development and Research"). Temperature (Constantz et al., 2001) and electrical resistance (Blasch et al., 2002) methods using small, inexpensive, waterproof sensors with integrated data storage also show promise for inference of streamflow timing in semiarid zones, especially in ephemeral channels with unstable beds.

Droughts can affect vast contiguous areas, leading to strong spatial correlation of low flows across a region. Because of this, periodic measurements of low flows can be used to extend information from streamgages to sites that are not continuously gaged. Potter (2001) examined the use of periodic measurements at ungaged sites to transform baseflow characteristics measured at the gage into estimates for the ungaged sites. He found that with as few as two periodic measurements per year, very good estimates of annual and long-term baseflow parameters (e.g., mean, median, lower decile) could be obtained. Such an approach during major droughts might be used to estimate baseflow parameters throughout the affected area, at scales much finer than those represented by the streamgage network. This information could help in understanding the geologic controls on the spatial variability of low flows during drought conditions. In addition, periodic measurements might also be made over many years at a few selected sites. This activity could provide valuable information on the interannual to interdecadal variations in baseflow response after an extreme drought.

The hydrodynamics of surface water-groundwater interaction can change dramatically during low flow when streams that normally receive groundwater discharge lose water if the adjacent water table drops below the stream surface water level. The transition from gaining to losing conditions can lead to significant biochemical processing of nutrients in the hyporheic zone. Similarly, during low flow, a streambed that was formerly covered by water is exposed, leading to discontinuous microhabitat zones for invertebrates and other fauna and flora, much like vernal pools and wetlands in the arid West. How these temporary microhabitats affect overall stream ecosystem health is not well understood. **Therefore, targeted intensive sampling of groundwater levels, geochemistry, and stream** morphology is needed during low flows as part of the NSIP's intensive monitoring for floods and droughts to improve understanding of these and other processes.

As the USGS intensive monitoring activities for 1993 Upper Mississippi River floods illustrate, the integration of flow and ancillary information can make significant contributions to river science for flow extremes. **Opportunities to collect, compile, and integrate ancillary information during major droughts also should be pursued.** For example, there is potential for the USGS to integrate low-flow measurements with soil moisture data from the U.S. Department of Agriculture (USDA) Soil Climate Analysis Network profiles (*http://www.wcc.nrcs.usda.gov/scan/*), AMERIFLUX long-term CO_2 flux measurement sites (*http://public.ornl.gov/ameriflux/Participants/Sites/Map/index.cfm*), and other local or state data. Such integration might even lead to tools that would assist predictive efforts on the effect of regional drought intensity on low flow.

Planning for Intensive Data Collection

The findings and conclusions of Parrett et al. (1993) after the 1993 floods on the Upper Mississippi River basin illustrate the potential contribution of intensive data collection during extreme hydrologic events to scientific study and understanding of river processes. This potential could be realized most effectively if the plans for intensive measurements were formulated to test scientific hypotheses related to flood and droughts processes. Even though it is impossible to anticipate where and when major events will occur, extensive pre-planning to identify scientific questions (requiring specific types of sampling and gathering of ancillary information to answer) and unique sites for scientific inquiry (where opportunistic measurements could be interpreted in a broader context) could significantly increase the information produced for scientific investigations.

Another consideration in the planning of data collection activities during major floods and droughts is estimation of flows at ungaged locations. There are opportunities to improve estimates of streamflow characteristics, particularly low flows, through regional analysis. **Plans for intensive data collection during major flood and drought events should be designed both to test scientific hypotheses on river processes, and to support regional analysis and estimation of streamflow information at ungaged sites.**

REGIONAL AND NATIONAL STREAMFLOW ASSESSMENTS

One of the most oft-cited reasons for having a national stream network with long-term records is the need to make assessments of streamflow characteristics across a region or the nation. Each gage by itself has an information content that increases as the record lengthens, which enables increasingly precise specification of the characteristics of streamflow at that location, such as the 100-year flood magnitude. When data from a set of gages in a region are assembled, the total information content is more than the sum of the parts, because regional patterns and coherence appear that are not visible in individual records.

Regional Flow Assessment

The use of streamgage observations from multiple sites in regional flow assessment provides valuable information for water resources decision making (NRC, 1992). The USGS is a leader in developing regional approaches to define streamflow characteristics such as the mean flow, flood peaks, or other percentiles of the flow distribution. Today, USGS districts routinely analyze observations from the streamgage network to provide regional regression equations for making flow estimates at ungaged sites. As the example in Chapter 4 for Texas (Slade, 2001) illustrates, regional flow estimation objectives are a key consideration in streamgage network design. Regional flow assessment traditionally focuses on statistical analysis of streamgage data. However, there are significant opportunities for integrating ancillary information in the study of regional flow processes. For example, the use of climate and weather data resources, as well as geographical information, can be integrated with streamflow information to examine and account for the effects of changing climate, land use, and other variables on regional flow statistics and flood frequencies.

Regional flow assessment can also contribute to a better understanding of hydrologic processes. As an example, a recent analysis of peak discharge records by O'Connor and Costa (2003) has helped to identify the factors controlling the largest floods observed in the United States. After pooling flood records at all sites and accounting for the dependence of flood discharge on drainage area, O'Connor and Costa (2004) identified the largest floods that have occurred in the United States and mapped their location. Figure 5-2 shows the location of the top 1 percent of flood peaks in the United States. The top 1 percent were found by plotting flood peaks versus drainage area; a threshold discharge curve was then used to define the top

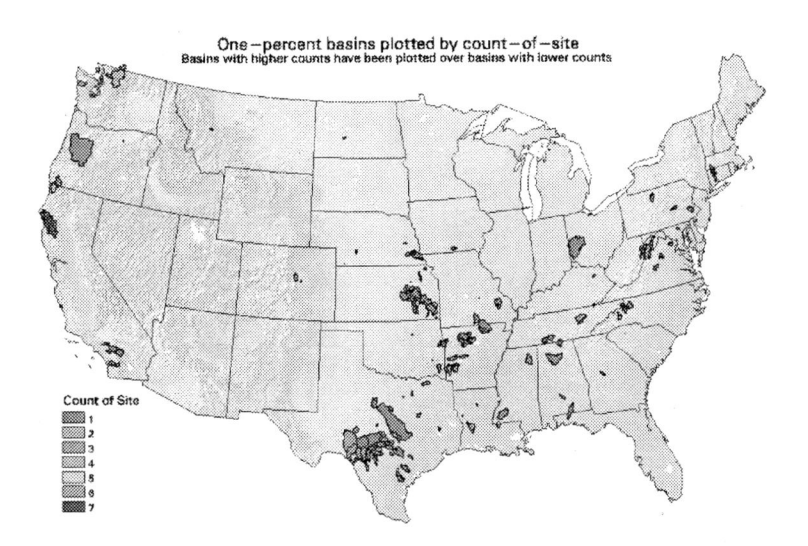

FIGURE 5-2 Drainage basins with the largest 1 percent of flood peaks recorded in the United States. SOURCE: J. Costa, USGS, written communication, March 2002.

events over the range of drainage areas. The results show that the location of the largest floods is not random throughout the United States. In fact, some basins had more than one flood among the top 1 percent. Some factors identified that make these areas prone to extreme flooding were the local topography, its interaction with atmospheric processes, and the proximity of the basin to atmospheric moisture sources. This and similar studies illustrate that regional hydrologic analysis of streamgage data has an important role in hydrologic science.

Long-Term Trends in Streamflow

One of the most important questions to be addressed in assessment of the streamflow network is, Are there long-term trends in streamflow? Such trends may be an indicator of the impact of climate change on water resources or the effects of human changes to the landscape. Using a subset of 395 streamgage records for the Hydroclimatic Data Network (HCDN), Lins and Slack (1999) examined trends in daily streamflow in the conterminous United States. The HCDN is a network constructed from existing USGS streamgages with watersheds that are relatively free of regulation, diversions, or land-use changes. Despite the popular perception that flood magnitudes are increasing, Lins and Slack (1999) found few significant trends in annual maximum flows across the United States. In contrast, significant and widespread trends were observed in lower flows, from the annual minimum to the median flow. These flows have increased across broad regions of the nation, except for the Pacific Northwest and the Southeast, where decreasing trends were observed.

In addition to the use of annual peak discharge (the annual series) (e.g., Lins and Slack, 1999), flood peaks as defined by the number of peaks above base (partial duration series) can also be valuable in flood frequency analysis and in the study of long-term trends in flooding. The two phenomena may be controlled by different processes. Traditionally, the USGS has reported both annual peak discharge and peaks above base. At present, however, these are not available on-line at the USGS web site, but they should be.

Questions regarding long-term trends in streamflow are relatively new and were probably not anticipated when USGS network streamgages were originally installed. However, with recent concerns over the potential effects of climate change on the water cycle, the availability of continuous long-term USGS streamgage records makes the study of trends possible. In addition to the study by Lins and Slack (1999), USGS streamgage records have been used to study long-term variability of monthly and annual flows throughout the United States (Chiew and McMahon, 1996; Lettenmaier et al., 1994; Lins and Michaels, 1994). These analyses have provided a valuable complement to investigations of the long-term variations in precipitation and precipitation extremes (Bradley, 1998; Karl and Knight, 1998; and Kunkel, 2003; among others). For instance, Karl and Knight (1998) observed significant, increasing trends in both precipitation and the proportion of total precipitation resulting from heavy precipitation events. The studies by Lins and Slack (1999) and others suggest that the hydrologic response to such changes has been an increase in low to moderate streamflows, but no discernible increase in flood magnitudes.

In addition to long-term trends, issues related to climatic variability and its impact on hydrology have emerged in recent decades. For instance, large-scale climate anomalies, such as the El Niño-Southern Oscillation and the Pacific Decadal Oscillation, are now known to affect streamflow variations over interannual to interdecadal time scales (e.g., Kahya and Dracup, 1993; Redmond and Koch, 1991; Sankarasubramanian and Lall, 2003). Increasingly, studies that integrate long-term streamflow and climate information are providing a hydroclimatic perspective on regional flow variations and extreme events. Examples of such investigations at the USGS include often-cited works on the impacts of large-scale climate forcing on snowmelt timing (Dettinger and Cayan, 1995) and the onset of spring (Cayan et al., 2001) in the western United States. Insights gained from hydroclimatological studies have also demonstrated the predictability of streamflow variations on a seasonal to interannual time scale, which may lead to better long-range streamflow forecasting (e.g., Hamlet and Lettenmaier, 1999). Additional studies of the linkages between streamflow, and climate and weather processes, are needed to advance scientific understanding of variations in the water cycle from local to global scales.

Overall, regional and national streamflow assessments are fundamental to NSIP and should be continued.

ENHANCED INFORMATION DELIVERY

The USGS is a leader in making its information and data easily accessible through the National Water Information System on the Internet (*http://waterdata.usgs.gov/nwis*), and these advances are especially compelling for real-time information.

Water Watch

The USGS Water Watch system (*http://water.usgs.gov/waterwatch/*) presents a map of streamflow conditions for the approximately 5000 streamgages whose data are acquired in real time (Figure 5-3). Each four hours, data are queried from the gages via the geostationary operational environmental satellites (GOES) system. For each gage and for each calendar day, the USGS has analyzed historical streamflow records to generate a percentage distribution of flow expected, and the actual flow is measured against these values to determine whether flow is above, below, or within normal flow conditions. A colored map of flow status is regenerated on the Internet every four hours with this information. Users can click on any station in this map and receive the "unit values," usually 15-minute streamflow and water-level data, for the past 30 days as a graph or as a data series. Given that it formerly took one to two years before daily mean streamflow data for gages were released, this real-time data delivery system is a great advance over past practices.

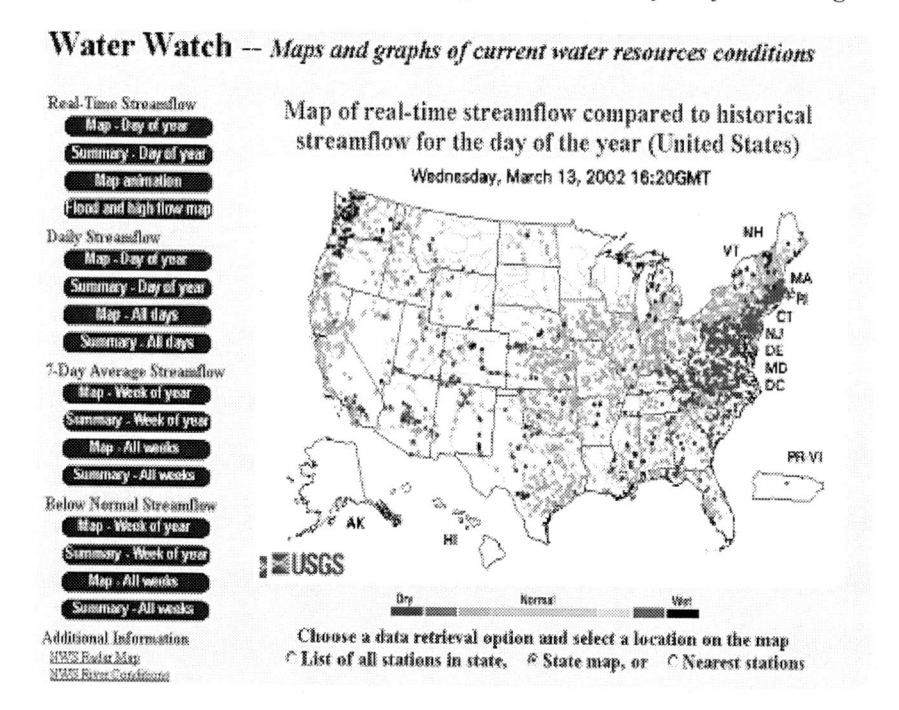

FIGURE 5-3 USGS Water Watch display for March 13, 2002, showing the regional drought in the Northeast. SOURCE: USGS (*http://water.usgs.gov/-waterwatch/*).

Real-Time Water Quality

The Kansas District of the USGS (*http://ks.water.usgs.gov*) has led the way in developing regression equations for real-time water quality display on the Internet (Christensen et al., 2000, 2002). In several streams in Kansas, the USGS measures, in real time, specific conductance, pH, water temperature, dissolved oxygen, turbidity, and total chlorophyll from sensors suspended in the water. Similar measurements are becoming routine at other water resources agencies, including publication of the observations on the Internet. However, the Kansas District work was innovative because it simultaneously collected periodic water samples and analyzed them for nutrients, bacteria, and other constituents of concern. Regression equations were then developed, and these equations were used to convert the real-time sensed variables into estimates with error bars of derived water quality variables.

This provided a continuous trace of water quality through time analogous to a streamflow hydrography. By combining estimated concentrations with flow, estimated constituent loads were also calculated, as illustrated in Figure 5-4 for fecal coliform bacteria. This is somewhat analogous to using a rating curve to convert measured water level into streamflow rate. Besides showing the estimated value, the resulting plots also show the range of uncertainty for these estimates. These data have a significant potential to inform Total Maximum Daily Load (TMDL) studies of water quality by quantifying the percentage of time that water quality standards are actually being met and the flow conditions under which they are not met. They also create an image of water quality and pollution loads varying through time with flow, which is not obtainable by viewing the results of periodic water quality sampling. By these means, water quality characterization at gage sites is placed on a continuous time basis as streamflow has been for many decades. The variability or extreme values of pollution concentration may in some cases be more critical for management than the mean concentration. For example, acidity loads to streams from abandoned underground coal mines may decrease stream pH to fish-killing levels only during low-flow conditions (Stoertz et al., 2001).

The provision of real-time water quality estimates analogous to those for streamflow is a very valuable adjunct to traditional streamflow information and, to the extent that resources permit, this capability should be expanded to other gages as quickly as possible.

Streamstats

In a pilot study initiated by the USGS Massachusetts District, a system called Streamstats has been developed to allow estimation of streamflow characteristics (mean, median, percentile values of the frequency distribution) at ungaged locations as a function of basin characteristics and regression equations (*http://ststdmamrl.er.usgs.gov/streamstats/*). When a user clicks on a desired location on the web-based map interface, Streamstats automatically determines the watershed draining to that location, applies the regression equations within the delineated watershed, and graphically displays the estimated streamflow values. This pilot study is being extended to several other states, and it is intended that Streamstats eventually will become a national system.

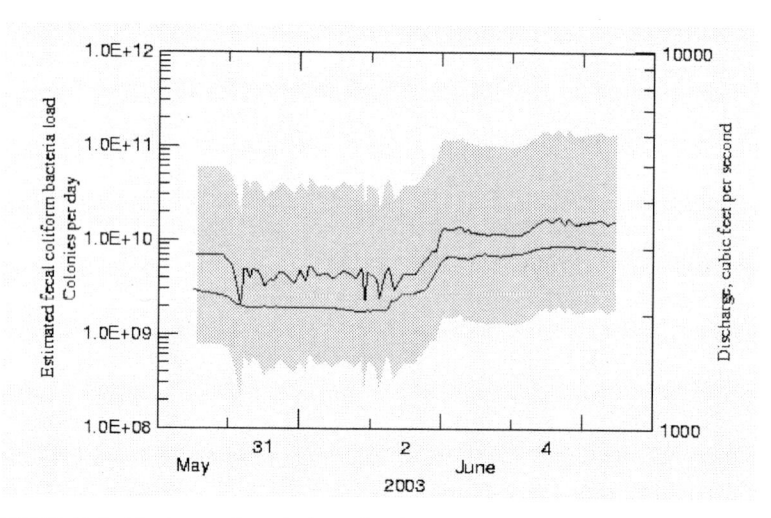

FIGURE 5-4 Estimated real-time fecal coliform bacteria load, with error bars shown, in the Kansas River at De Soto, Kansas. Discharge is shown for comparison. SOURCE: USGS (*http://ks.water.usgs.gov/Kansas/rtqw/index.shtml*).

Streamflow Information Products

Two traditional roles of the USGS have been the measurement and publication of historical daily mean streamflow data and streamflow statistics. Increasingly, provision of real-time data is occurring at streamgages through Water Watch. Also, a capacity is being developed to estimate streamflow statistics at ungaged sites with Streamstats. One can thus think about streamflow information in terms of location, such as at a streamgage or an ungaged site anywhere on the stream, and in terms on the time scale of the product, such as real-time data, daily summaries of historical observations, or statistical characteristics of the flow based on historical data. This conceptualization is illustrated in Figure 5-5, where the size of the filled circles illustrates the degree to which products are currently available at different locations. In a more complete system, shown by the open circles, a user would be able to estimate historical and real-time streamflow at ungaged locations in an analogous manner to stream statistics.

Another streamflow information product that would be useful in science and engineering applications is finer-resolution discharge observations. At present, real-time data are published as unit values, that is, for each interval within the day that the data were measured. However, only the daily

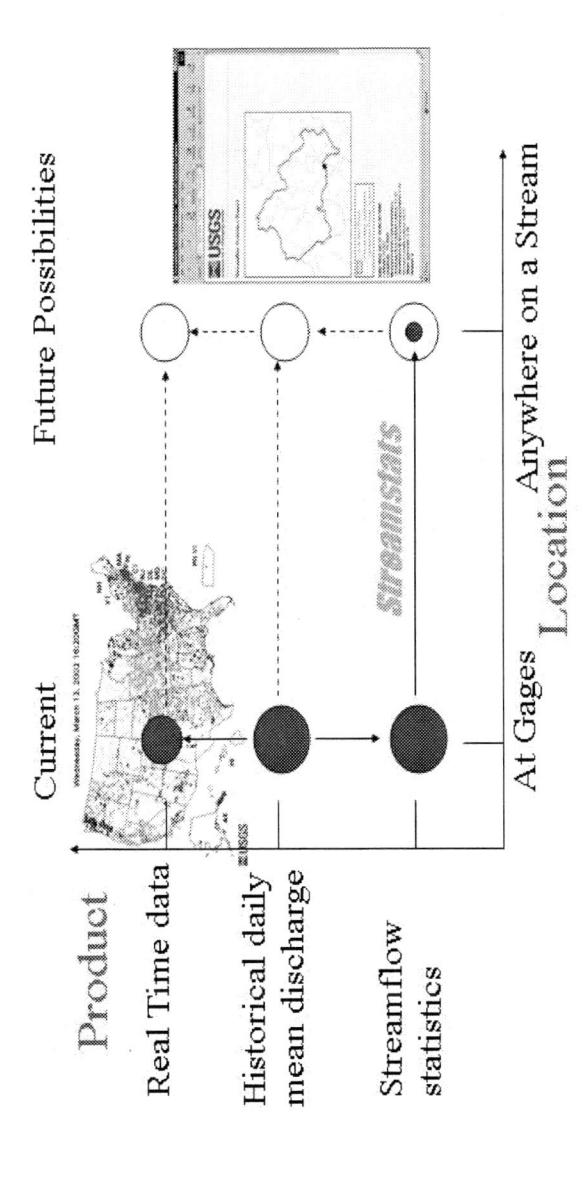

FIGURE 5-5 Streamflow information products and locations at which they are available. Filled circles represent the current capability, with the size of the circle representing the availability of data. Open circles represent future capabilities.

mean values are published as historical data in the National Water Information System (NWIS). **The USGS should develop a system for publishing the unit value data so that historical streamflow data can be obtained for intervals of less than one day.** These data would be of great value, for example in flood estimation studies on small basins where the duration of flood events is much less than one day.

Flood Inundation Simulation Using Two-Dimensional Flow Modeling

There is a significant public interest in real-time flood inundation mapping, especially if presented on the Internet or on television so that people can avoid flooded areas. Jones et al. (2002) have presented a pilot study of near-real-time flood simulation and Internet delivery of flood inundation maps in the Snoqualmie River, Washington. In this simulation, the input flows were generated by the National Weather Service River Forecast Center, and the inundation surface was generated by a flood model called TrimR2D that can reproduce backwater effects resulting in water in otherwise unflooded side channels draining into the main river. The resulting map was presented using an Internet map server. Other organizations are also working on real-time flood inundation mapping, including the Hydrologic Engineering Center (HEC) of the U.S. Army Corps of Engineers, which has created a Corps Water Management System that ingests real-time rainfall and streamflow information and computes flows, water surface elevations, and flood maps using HEC models embedded in the system.

Creating inundation maps over large stream networks requires having good measurements of the stream cross section along a river profile. Much of this information is stored in regression equations relating stream width and depth to drainage area and other variables. Currently available digital terrain data can be used to describe the inundation area in the floodplain. What is missing is sufficient detail about the geometry of the stream channel to support accurate flood inundation mapping. **The USGS should develop the capability to estimate stream channel characteristics at ungaged locations along significant rivers and streams.**

METHODS DEVELOPMENT AND RESEARCH

Methods development and research refers to advances in techniques for direct measurement of streamflow. For more than a hundred years, current meters have been the standard for making direct discharge measurements. Although a single discharge measurement can take an hour or more for large rivers, the technique is well documented (Buchanan and Somers, 1969) and accurate (Pelletier, 1988; Sauer and Meyer, 1992). In recent years, acoustic Doppler current profiler (ADCP) devices have been introduced for discharge measurements on larger rivers; all USGS districts have now been equipped with at least one of these devices. ADCP uses an immersed acoustic probe to measure velocity profiles from a floating platform on the water surface. Some advantages of using ADCPs are that measurements can be made much more rapidly than with current meters (i.e., minutes rather than an hour) and the device produces detailed information on velocity profiles, which is used directly for discharge estimation. A disadvantage of the ADCP is that it is unable to measure velocities near the water surface or the river's bed. This limitation restricts its use to relatively large rivers. There are other limitations of these conventional approaches that affect USGS streamgaging operations. For example, making measurements with current meters or ADCP requires contact with the flow. This can be hazardous to people or equipment, especially during a flood measurement.

Because of the limitations of existing measurement devices, the USGS has formed the HYDRO21 Committee to investigate and test new approaches to discharge measurement. The focus of the committee's work has been on remotely sensed, non-contact methods for gaging streams (Melcher et al., 1999). Unlike conventional techniques, current non-contact technologies are only capable of measuring surface velocities. Therefore, an assumption regarding the velocity profile, or complex hydraulic analysis, is needed to estimate discharge from surface velocity measurements. As with conventional approaches, discharge estimation also requires a measurement of the channel cross section. Promising techniques include Doppler radar and visible imagery techniques for surface velocity measurement and ground penetrating radar (GPR) and light detection and ranging (lidar) for channel bathymetry measurement.

Doppler radars send out electromagnetic pulses, which are reflected back to a sensor by periodic waves on the water's surface through a process known as Bragg scattering (Plant, 1990). The surface waves on a river are generated by wind, river turbulence, floating debris, and other processes. Both monostatic (an integrated transmitter and receiver) and bistatic (separate transmitter and receiver) sensors have been investigated. Since radars can only detect motion in the direction of the beam's path, the flow direction is assumed in order to estimate surface velocity vectors. Visible imagery techniques use digital images of the flow surface to detect surface motion. A cross-sectional technique, known as particle image velocimetry

(PIV; Adrian, 1984), is used to detect motion from image pairs. Although PIV is a standard technique for laboratory flow measurement, it has only recently been explored for measuring river flows (Creutin et al., 2003). Because it uses visible images, measurement can be made only in daylight, and there must been visible motion at the surface, from debris, eddies, or waves. GPR is used extensively to map the subsurface in geophysical applications. GPR measurement of channel bathymetry uses low frequency band wavelengths (60 to 300 MHz) to distinguish between air, water, and sediment boundaries. The radar must be suspended in close proximity to the water surface for measurement. Since a GPR signal is strongly attenuated in high sediment loads, measurements cannot be made when the turbidity is high. In contrast, lidar uses laser pulses to measure air-water-sediment boundaries. Lidar can make measurements from higher altitudes (a few hundred meters), but its resolution would average depths over relative large areas (a few square meters).

The HYDRO21 Committee has tested components of such non-contact devices in several "proof-of-concept" experiments. Spicer et al. (1997) used a GPR to measure cross sections of four streams near Mount Saint Helens, Washington. By suspending the GPR from a bridge or a cableway, they found that they could reliably create a plot of the streambed cross sections. Costa et al. (2000) combined GPR with Doppler radar to make a discharge measurement on the Skagit River, Washington. A suspended Mala Geoscience GPR measured water depths, and the University of Washington X-band Doppler radar measured surface velocities from the river's bank. Depth-averaged velocities were estimated by multiplying the surface velocity by 0.85 (assuming a parabolic velocity profile) and integrated with the cross-section information to estimate discharge. The resulting discharge estimate was remarkably similar to that based on current meter measurements (less than a 0.2 percent difference).

More recently, Melcher et al. (2002) made discharge measurements on the Cowlitz River at Castle Rock, Washington, from a helicopter using Doppler radar and GPR. The helicopter hovered 3-5 m above the water surface during the experiment, and measured surface waves induced in part by the propeller wash of the helicopter (see Figure 5-6). Depth-averaged velocities were estimated from surface velocities every 3 m across the river; the estimates were multiplied by the corresponding depths and summed across the river to obtain the discharge. The results for mean velocity and depth were within 2 percent of those obtained by a simultaneous sounding weight and current meter measurement, and the radar-estimated discharge was within 0.4 percent of the current meter discharge.

Streamflow Information 117

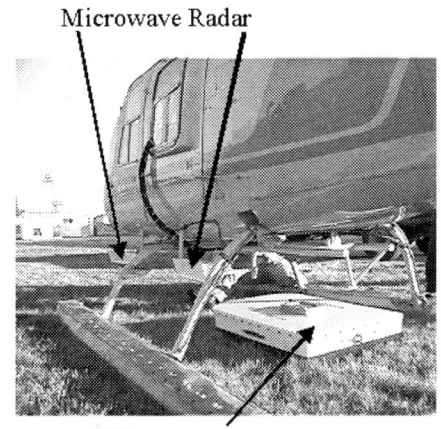

Microwave Radar

GPR antenna

FIGURE 5-6 Helicopter experiments to measure discharge. SOURCE: John Costa, USGS, written communication, March 2002.

Other investigators have examined river discharge measurement using imagery techniques. For example, Bradley et al. (2002) used a video camera to visualize the flow seeded with tracers on Clear Creek, Iowa. Surface velocities were then estimated using particle image velocimetry (PIV) with 60 seconds of images. A hydraulic model based on kinematic principles (conservation of mass) was used to derive three-dimensional flow field for discharge estimation. The discharge estimated by this approach was within 1 percent of the current meter measurements.

The near-term goal of the HYDRO21 Committee's work has been to develop the "gaging station of the future" (Figure 5-7). For example, a future gaging station might consist of a permanently installed pulsed Doppler radar to measure velocity continuously, a GPR to make periodic measurements of channel bathymetry, and a satellite system to transmit data in real time. Still, the committee also envisions tailoring techniques to unique applications, such as those required to make intensive measurements at ungaged sites during floods and droughts. These other applications might use technologies such as video image analysis for discharge estimation or handheld radar guns for spot measurement of surface velocities, increased use of lidar for floodplain mapping or enhanced forms of lidar that can penetrate water for mapping stream bathymetry, and the remote sensing of water surfaces and areas of flow inundation using land-, aircraft-, or space-based sensors.

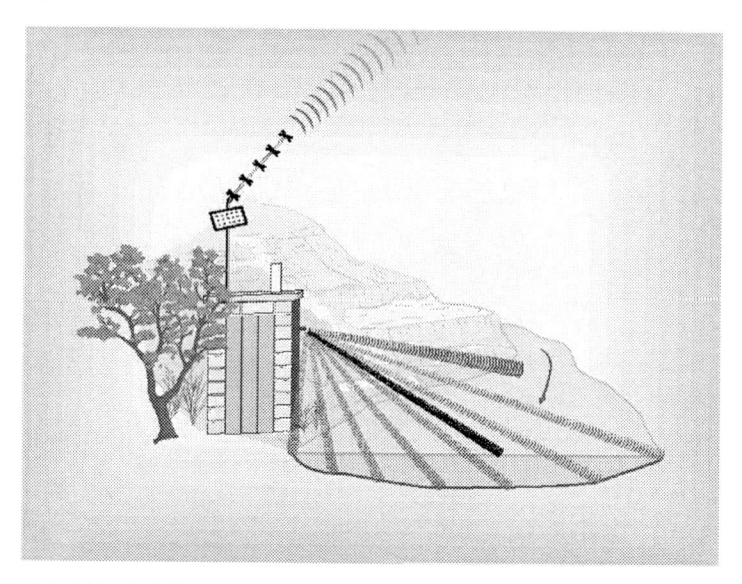

FIGURE 5-7 USGS streamgaging station of the future. SOURCE: U.S. Geological Survey (1999, p. 13).

In all of the technologies described above, a very careful evaluation of these techniques before and after they become operational is critical. The advantage of the relative lack of advancement in streamgaging technology in the last century is the consistency and comparability of data over this time. Even when a newer technique is proven superior over an older one, care must be taken to ensure that technique-based nonstationarities in the rich, long-term historical records of streamgage measurements are not created.

With due care in ensuring comparability between traditional streamgaging data and new technologies, the USGS is encouraged to continue aggressively pursuing these technologies for measurement of streamflow and related parameters with a view to accelerating the implementation of time- and labor-saving flow measurement techniques, and continuous water quality monitoring, as soon as practicable.

SUMMARY

In general, the four other components of the NSIP that complement the streamgaging network—intensive data collection during major floods and droughts, assessments of streamflow characteristics, streamflow information delivery to customers, and methods development and research—are well conceived and appropriate to the USGS. The spatial scale and risks of hydrologic extremes (e.g., floods and droughts) are areas deserving the attention that the USGS proposes in the NSIP. In particular, targeted intensive sampling of groundwater levels, geochemistry, and stream morphology are needed during low flows as part of NSIP's program of intensive monitoring for floods and droughts to improve our understanding of these and other processes. This information should be integrated with ancillary data such as soil moisture and CO_2 flux data as appropriate. Plans for intensive data collection during major floods and drought events should be designed both to test scientific hypotheses on river processes, and to support regional analysis and estimation of streamflow information at ungaged sites. The USGS should further refine its information delivery strategy to include on-line, value-added products, such as flood simulations and water supply and water quality projections under various development scenarios. The USGS should also disseminate more types of data, including historical data (requiring rescue of older paper format data), cross sections, velocity profiles, unit discharge values, and opportunistic data (e.g., crest stage data and slope-area data from flood studies). This is likely to require changes in the data management system to accommodate these various data types and formats.

Many research opportunities that should be pursued, including the following:

- Development and use of a portfolio of data collection tools in addition to the fixed, permanent stations, such as acoustic Doppler current profilers to measure stream velocity and channel resistance
- Real-time water quality estimates analogous to those for streamflow
- Measurement of streamflow at ungaged sites during high- and low-flow conditions using mobile units
- Spatial and temporal trends in streamflow, especially with respect to floods and droughts

6
Contributions of NSIP to River Science

Rivers do more than simply convey the water, sediment, and dissolved components from the watersheds they drain. Streams and rivers have distinctive channel characteristics that are the product of the flow regime's capacity to transport the sediment supplied to the channel. The interaction between water, sediment, and in some instances, large woody debris creates many aquatic and subsurface habitats for the diversity of riverine life, from microorganisms to insects to fish to riparian trees. Groundwater delivered to streams and surface water in the stream can be biogeochemically transformed in subsurface hyporheic zones beneath and around the streams. An understanding of the functioning of the integrated hydrological, geomorphic, and biological processes in rivers is a fundamental goal of river science, and it requires information on streamflow, water quality, and sediment load. This understanding is complicated because of the substantial imprint of human activities on river systems, activities that can greatly modify geochemical, physical, and biological processes. These processes are sensitive to land-use change and climate change; therefore, one key way that in which the National Streamflow Information Program (NSIP) can support river science is by providing information on how human activities influence key processes that alter a river system relative to some minimally disturbed "reference" conditions (such as might be provided by the sentinel watershed element of NSIP; see Chapter 3).

Streams and rivers also provide numerous goods and services to society, such as water supply, recreation, hydropower generation, food production, and aesthetic values. Demands for these goods and services are increasing as population grows and as concerns about recurrent drought and climate change increase (Postel et al., 1996; Vörösmarty et al., 2000). At the same time, societal interest in maintaining the ecological sustainability of these flowing water ecosystems is growing, leading to potential conflicts between perceived human and ecosystem needs for fresh water (Baron et al., 2002; Naiman et al., 2002). Potentially conflicting demands can be expected to increase into the future due to pressures of population growth and climate change, which will only intensify society's need for better scientific information and understanding required to manage the nation's freshwater resources (Poff et al., 2003).

As an example, the closure of Glen Canyon Dam in 1963 changed the magnitude, timing, and temperature of streamflow and reduced sediment inputs into the Grand Canyon segment of the Colorado River. This has impacted the number and sizes of sandbars which are used by river runners and form the habitat for native fish. An experimental flood was released from Glen Canyon Dam in 1996 in an effort to rebuild sandbars and evaluate the potential for controlled flooding as a management tool (Webb et al., 1999). Scientific understanding of the interaction of geomorphologic, hydrologic, and biologic processes within rivers is needed to support this kind of management. The U.S. Geological Survey (USGS) has a critical role to play, through streamgaging and more comprehensive river process studies, in water resources prediction and in support of river management in the coming decades.

The committee was asked to address the following statement of task: How does the National Streamflow Information Program support river science, and can it support an integrated river science program in addition to its operational objectives? In that context, the purpose of this chapter is to briefly review opportunities in river science provided by the existence of the NSIP and to identify some additional requirements for streamflow information and dissemination to support river science.

RIVER SCIENCE OPPORTUNITIES CREATED BY THE NSIP

The term *river science* as used in this report is a largely interdisciplinary field that includes surface and groundwater hydrology, fluvial geomorphology, and various subdisciplines of biology (e.g., biogeochemistry, riparian ecology, aquatic ecology). The USGS is in a unique position to play a very important, leading role in guiding the development of a river science that can support society's broader concerns about river sustainability and management. The growing need for scientific information on rivers affords the opportunity for the USGS to define and explain how the development of a river science program represents a desirable societal investment. The USGS has already demonstrated its role in providing high-quality scientific information in a number of high-profile river management contexts including, for example, the Missouri River (Auble and Scott, 1998) and the Glen Canyon controlled flood on the Colorado River discussed above.

The opportunities for involvement of the USGS in river science, however, are significantly greater than its current role. The primary service provided by the USGS in enhancing river science would be to collect and provide the information needed to advance scientific understanding of the natural biophysical processes that define river systems and to build scientific capacity to predict how human alterations affect these processes for streams and rivers across the nation. Secondarily, the USGS should be in a position to provide unbiased scientific expertise in river science as requested by the public in the management of rivers.

Figure 6-1 illustrates a simplified view of the core USGS disciplines that contribute information and data fundamental to river science. This view of river science is inherently interdisciplinary and envisions integrated interaction among component disciplines, as well as interactions with related disciplines (e.g., hydroclimatology, geology). Most scientific studies on rivers conducted to answer federal or state questions require that data be acquired in addition to those normally collected at streamgage sites. However, the USGS's NSIP currently provides, and will continue to provide, the basic infrastructure for these studies.

Streamflow Information Needs for Geomorphic Studies

The USGS has been a leader in the development of scientific fields that are anchors of its river science program. One of the strongest examples of this is the field of fluvial geomorphology, which has developed with strong support of USGS streamflow information and strong scientific leadership from within the USGS. For example, the term "hydraulic geometry" refers to the changes in hydraulic variables (width, depth, velocity) that increase to accommodate increases in discharge either at a gaged site or at successive locations in the downstream direction. The seminal paper on hydraulic geometry is Leopold and Maddock (1953). This research was made possible by the existence of streamflow and channel morphology measurements at USGS gaging stations.

Information on the hydraulic geometry of rivers has been published in various regions of the world. Surprisingly, the USGS and other groups have not published hydraulic geometry relationships (either at a station) or downstream for hydroclimatic regions of the United States. A consequence

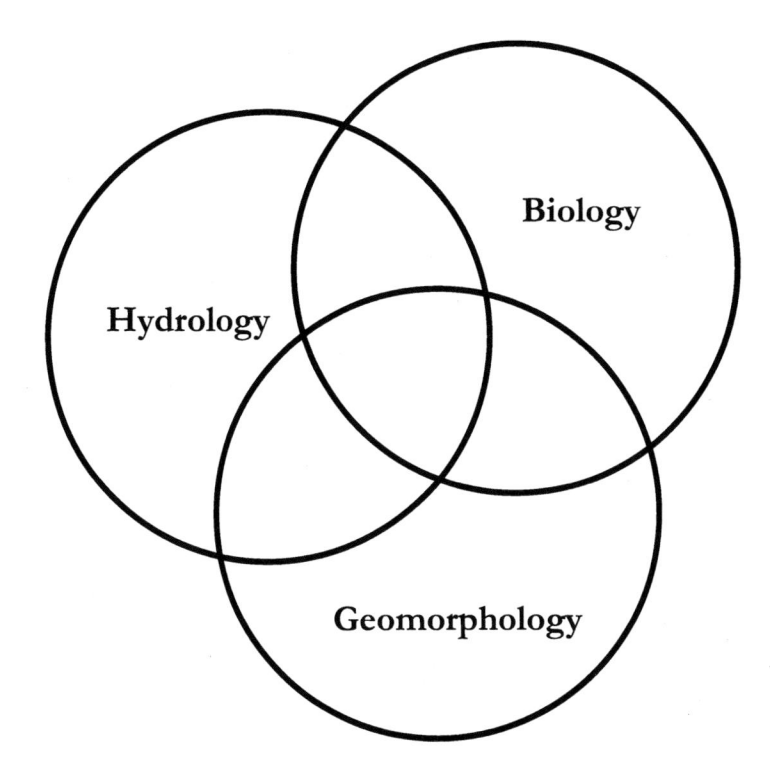

FIGURE 6-1 Venn diagram illustrating primary disciplines contributing to river science, an interdisciplinary endeavor represented by overlap among the disciplines.

of this is that many research projects that require channel hydraulic geometry use either "average" hydraulic geometry relationships, which are often the data from Leopold and Maddock (1953), or stream classification schemes (e.g., Rosgen, 1994), which are most appropriately applied in situations lacking high-quality data.

The USGS has begun to publish data from the individual streamgagings made at each active gage site (*http://waterdata.usgs.gov/nwis/sw*), which are essential for the evaluation of hydraulic geometry relationships. A limitation of this data source is that USGS gaging stations are chosen to have particular channel characteristics, such as the existence of a control section that will ensure a unique rating curve. The channel characteristics of streamgage locations may thus not be representative of randomly selected locations at any point along the entire length of a stream or river.

Geomorphic studies also require information that is sometimes, but not always, collected at streamgaging stations. These data include stream gradient, bed grain sizes, suspended sediment transport, and bedload transport. Stream gradient and bed grain sizes are essential for evaluation of bed mobility or sediment transport capability of a stream. Stream gradient is required to estimate local or reach-averaged stream power and shear stress. Further, flow resistance of a river can be calculated if stream velocity, hydraulic radius, and stream gradient are known. Flow resistance is a parameter that is used in all hydraulic models, including flood routing and flood inundation. USGS streamflow data provide an important data set that can be used to evaluate flow resistance, provided stream gradient is known.

Grain size information is also essential geomorphic information that is required both for geomorphic studies of channel morphology, sediment transport, and channel changes and for many ecological studies as well. Grain size information can be used to evaluate the mobility of bed sediment in rivers. At each USGS gage site and other reaches, the mobility of bed sediment could be evaluated if grain size distributions and the stream gradient data were measured in addition to the existing streamflow data.

Sediment transport data are expensive and difficult to collect, but because sediment load is an independent variable in stream systems and is highly variable spatially, these data must be collected from a range of watersheds. The USGS has a collection of suspended sediment data on streams that can be used to develop suspended sediment rating curves and loads. New technologies also hold promise for enhancement of data collection programs. For example, acoustic Doppler current meter data provide information on the variation of velocity with depth. These data can be used to evaluate roughness heights, local shear stress values, mixing lengths, and cross-channel shear stress distributions. These data provide a real opportunity to significantly enhance the hydraulics and sediment transport program at the USGS.

Streamflow Information Needs for Biological Studies

The hydraulic characteristics of river channels serve as determinants of many ecological processes and patterns in streams, through both direct effects on organisms and indirect effects mediated by factors such as sediment and wood transport and storage ("habitat"). Temporal variation in streamflow creates dynamic hydraulic variation that can reconfigure channel morphology and habitat for aquatic organisms and thus influence many ecological processes, both within the channel and on adjacent floodplains that experience inundation.

In the past decade or so, the general importance of hydrologically generated "disturbance" has become widely recognized in river ecology (e.g., Junk et al., 1989; Poff et al., 1997; Resh et al., 1988). Streams and rivers are naturally dynamic systems, due to frequent fluctuations in flow conditions. The occurrence of extreme events (floods, droughts) in particular is ecologically significant in that they typically "reset" ecosystems by creating sets of conditions that benefit early successional species and thus maintain high diversity. In other words, flow variation helps establish a "habitat template" that regulates many ecological process rates and influences the distributions and abundances of species (Poff and Ward, 1990; Schlosser, 1987; Townsend, 1989; Townsend and Hildrew, 1994). Several good reviews of this topic are available (Bunn and Arthington, 2002; Gasith and Resh, 1999; Poff et al. 1997). Indeed, there is now great interest in using long-term hydrologic data from USGS streamgages to characterize hydrologic disturbance regimes both within individual streams and among streams in a comparative fashion that allows for classification of regime types and enhanced ability to predict ecological responses to human alterations.

The USGS gage network has been instrumental in the progress of "hydroecology" in the last decade. For example, regional flow regime classifications have been constructed based on hydrological variables that are explicitly relevant to ecological processes in streams and rivers. These hydroecological classifications emphasize the patterning of flow variability at multiple time scales, described in terms of frequency, magnitude, duration, timing, and rate of change of flow events with ecological relevance (Olden and Poff, 2003). Computer software tools are now available and widely used to assist in codifying this approach (Richter et al., 1996). Several hydroecological classifications have been developed around the world for unregulated streams in the United States (Poff, 1996; Poff and Ward, 1989), Australia (Hughes and James, 1989), and New Zealand (Clausen and Biggs, 2000). An example of a U.S. classification based on more than 800 streamgages is shown in Figure 6-2.

Streamflow information from USGS gages is also critical for many site-specific hydroecological investigations. For example, Friedman and Auble (1999) used long-term streamflow records and dendrochronology to quantify the survival patterns for box elder stress along a gradient of flood inundation and shear stress in a section of the Black Canyon of the Gunnison National Park. They combined empirically derived relations between flow and tree mortality with a hydraulic model of the Gunnison River bottomland to generate Figure 6-3, a mortality response surface expressed in terms of key streamflow variables. Such a model provides park managers a tool for determining how upstream reservoir operations might be manipulated

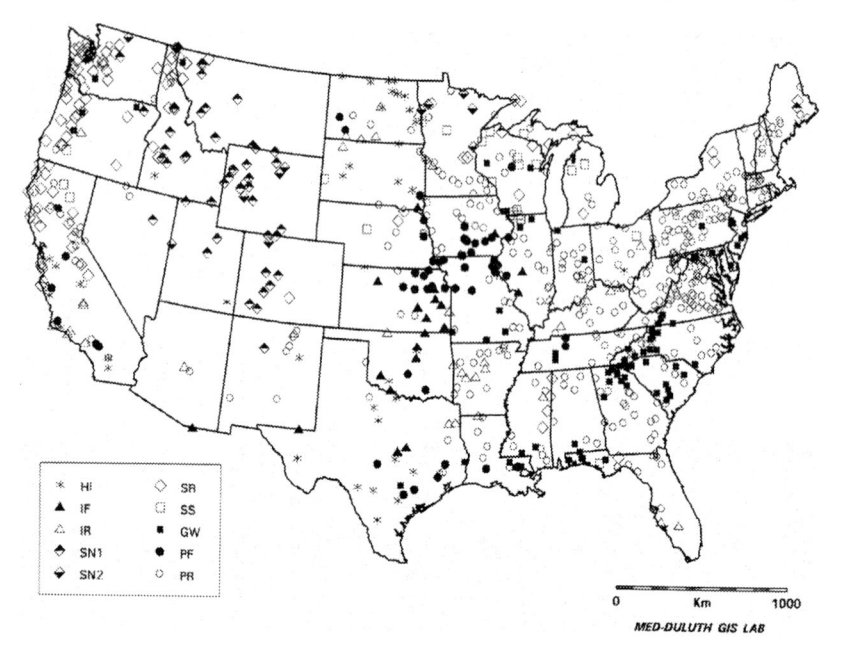

FIGURE 6-2 Ecohydrologic classification of 816 unregulated streams in the United States based on long-term daily streamflow data from USGS gaging stations. NOTE: Abbreviations refer to 10 streamflow "types" identified from cluster analysis based on 11 hydrologic variables: HI = harsh intermittent; IF = intermittent flashy; IR = intermittent runoff; SN1 = snowmelt 1; SN2 = snowmelt 2; SR = snow + rain; SS = superstable groundwater; GW = groundwater; PF = perennial flashy; PR = perennial runoff. SOURCE: Poff (1996).

to control the growth of box elder in the national park. As another example, innumerable studies are conducted by state and federal agencies throughout the United States to evaluate minimum instream flows for fish using techniques of quantifying time series of hydraulic habitat conditions, and these almost always require the availability of high-quality flow data (IFC, 2002).

Ecological studies, therefore, require information on the amount, flow rate, and timing of streamflow that regulates many of the ecological functions of the stream. Although many of the data collected at NSIP gages are appropriate for ecological studies, there is often insufficient information available for small streams. Geomorphic data are also required for many ecological studies, and therefore the data needs described above are also re-

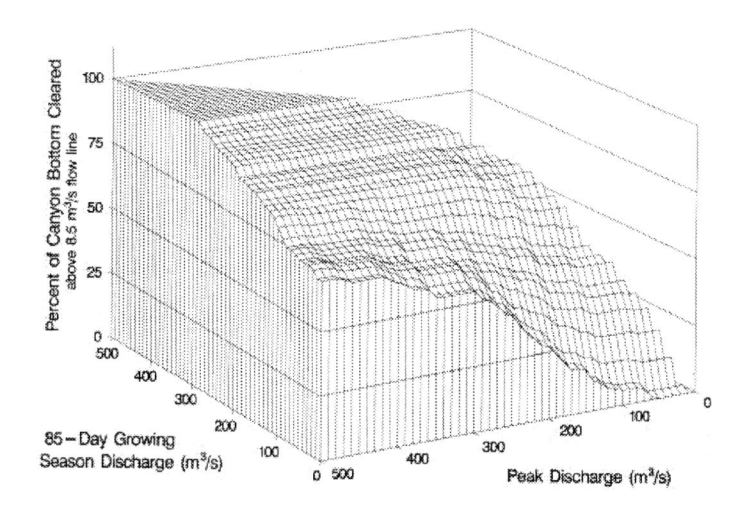

FIGURE 6-3 Mortality response surface for box elder trees as a function of flood magnitude and seasonal inundation. SOURCE: Friedman and Auble (1999).

quired for many ecological studies, amplifying the need for the dissemination of data that are currently not readily available. Ecological studies also require information at ungaged locations, indicating the need for development of streamflow estimation and geomorphic estimation procedures.

Streamflow Information Needs for Surface Water-Groundwater Interaction Studies

The hyporheic zone is the subsurface interface between stream water and the groundwater interacting with it (Figure 6-4). Groundwater can discharge to streams and maintain base flow and in turn, be recharged by streams (Figure 6-5). Groundwater flow patterns also can be influenced by stream gradient and geomorphology, and anthropogenic influences such as local pumping and water use.

The three-dimensional extent of the hyporheic zone and its hydrodynamics are related to overall streamflow dynamics (e.g., Battin, 1999; Jones and Mulholland, 2000), and within the hyporheic zones, focused groundwater discharge through macropores or other highly permeable zone can lead to unique biological habitats. The hyporheic zone is important both in

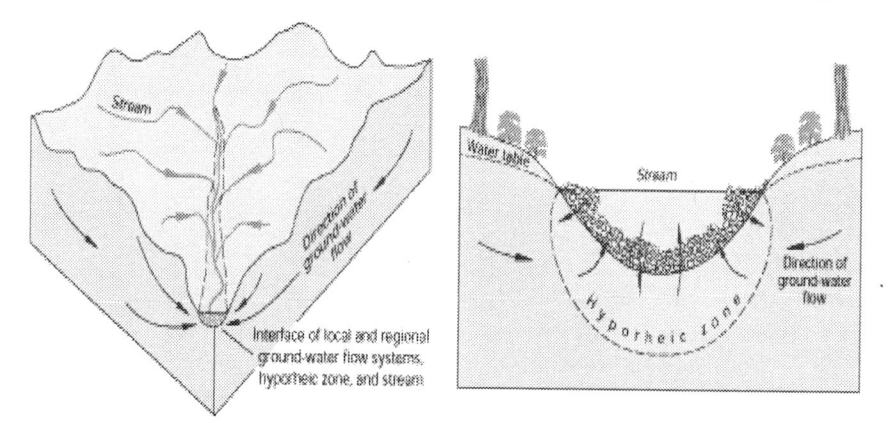

128　　　　　　　　　　　*Assessing the National Streamflow Information Program*

FIGURE 6-4　The hyporheic zone. Note the "envelope" of water under the stream that is active with respect to water fluxes and mixing and geochemical processes. SOURCE: Winter et al. (1998; http://water.usgs.gov/pubs/circ/circ1139/htdocs/natural_processes_of_ground.htm).

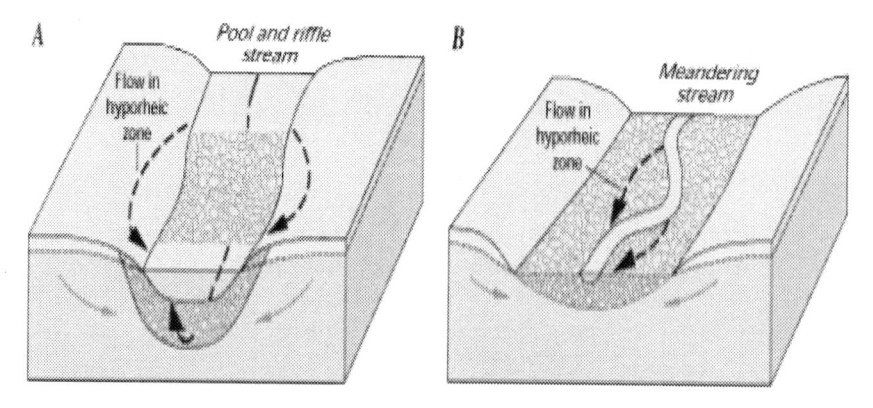

FIGURE 6-5　Water cycling between the groundwater system and streams (a) where pools and riffles create abrupt changes in the slope of the streambed and (b) at stream meanders. SOURCE: Winter et al. (1998; http://water.usgs.gov/pubs/circ/circ1139/htdocs/natural_processes_of_ground.htm).

terms of biogeochemical transformations (e.g., Cirmo and McDonald, 1997; Grimm and Fisher, 1984; Harvey and Fuller, 1998; Hill et al., 1998; Hinkle et al., 2001; Triska et al., 1993) and as habitat for a wide variety of organisms (e.g., Hendricks, 1993). Fish and other biota are often highly sensitive to temperature and stream water quality at stream margins, which are partly controlled by the proportions of groundwater entering and leaving the stream.

Spatial and temporal gradients in dissolved oxygen, dissolved organic matter, and solutes can be profound in the hyporheic zone, which is where most nutrients and, logically, anthropogenic contamination to streams is processed (e.g., Harvey and Fuller, 1998; Jones and Mulholland, 2000; Nagorski and Moore, 1999; Schindler and Krabbenhoft, 1998; Winter et al., 1998). Geochemical changes in the hyporheic zone are coupled to microbiological processes (e.g., Hendricks, 1993).

The hyporheic zone controls not only transverse geochemical processes at the surface water-groundwater interface, but sometimes even longitudinal geochemical processes downstream (e.g., Wörman et al., 2002). The hyporheic zone in many places is the fundamental driver for geochemical processing and even weathering in watersheds over a wide range of hydrogeologic settings.

The clear linkage between the hyporheic zone and biological diversity and habitat there has made the study of hyporheic processes one of the richest areas for multidisciplinary research. Indeed, the hyporheic zone is now considered a distinctive ecotone (e.g., Vervier et al., 1992) wherein new instrumentation is being developed to better describe subtle and transient changes in pore-water chemistry and hydraulics (e.g., Duff et al., 1998; Geist et al., 1998).

It stands to reason that part of the scope of the NSIP could be tied to monitoring hydraulic and other parameters related to interaction in the hyporheic zone. For example, inexpensive pressure transducers or thermistors could be installed adjacent to small headwater streams to monitor directional changes in groundwater flow relative to the stream and the extent to which periodic flooding affects the fundamental hydraulics associated with floodplains. The data output from these devices could be sampled remotely along with stream stage. At the very least, the NSIP could provide reconnaissance data to help biological and hydrologic scientists determine where best to focus more detailed studies designed to determine the fate and transport of nutrients and anthropogenic contamination to streams.

Interdisciplinary research in hydrology, geomorphology, biology, and groundwater-surface water interaction is also being done at experimental watersheds operated by other federal agencies, such at the U.S. Forest Service and the Agricultural Research Service. Close coordination with the efforts of these agencies and the academic communities that work at these sites is, of course, desirable.

INFORMATION NEEDS FOR RIVER SCIENCE

There are two overarching information needs for river science. First, information must be generated that will promote an integrated, process-based understanding of hydrologic-geomorphic-biological linkages. A good example is channel geometry and bed material composition. These are critical information needs to evaluate the hydraulic characteristics of a river reach or even a whole network. They allow models of sediment and hydrologic routing to be used. The temporal and spatial characteristics of this material routing are of central importance to understanding many key ecological processes that influence ecosystem resilience and provide ecosystem goods and services.

Second, models should be developed that allow point information to be distributed spatially, both within the gaged watershed and into ungaged watersheds. Such interpolations will allow process-based models to be extended spatially. Equally as importantly, they will also allow biophysical comparisons between watersheds to be drawn that support classifications for research and management. Essentially, they provide a foundation for establishing the degree to which biophysical and geochemical processes have been altered by human activities and thus what kinds of management and regulatory actions might be required.

Both of these needs can be met only if there is an extensive streamflow gaging network that has representative coverage of the range of climatic and watershed characteristics across the United States. This section reviews the streamflow information available at NSIP gages and its suitability for the needs of the river science community as described above.

Streamflow Information Issues for River Science at NSIP Gages

As described earlier in this report, the streamflow information that is collected at gaging stations provides a wealth of information that can be used to evaluate the frequency and magnitude of floods that shape the channel and riparian vegetation. Flow duration information is also available for active streamgages and is used for geomorphic and ecological studies. Some problems with using these data for river science purposes are reviewed here:

- **Nonstationarity.** Hydrologic time series are the primary source of information used to construct water budgets at any particular spatial scale. Projections of future water yield or demand for human and ecological needs are based on these time series. These hydrological time series are usually assumed to be stationary, as in the Hydroclimatic Data Network. The robustness of this assumption has to be rigorously evaluated given the change in climate across the United States during the twentieth century. When the streamgage network was first established, it was thought that streamgages could have a limited lifetime to establish the characteristics of the flood frequency regime. Land-use change also influences hydrologic flux and therefore represents another source of non-stationarity in longterm hydrologic records. Even in watersheds minimally influenced by humans, vegetative cover can change naturally in response to climatic variations. For example, the precipitation regime can control the extent of vegetative cover in a watershed and the probability of fire that can eliminate established vegetation. Comprehensive integrated analyses of hydrologic-climatic-landscape linkages are needed to assess nonstationarity introduced by climate variations or land cover evolution. Such analyses provide information about the streamflow variability that is essential for analyzing ecosystem and geomorphic processes critical to river science.
- **Estimation of extreme events.** The USGS gaging network performs well in monitoring and reporting moderate- to high-flow conditions on the nation's streams and rivers. By comparison, low-flow measurements can be relatively poor because gages are better suited to measuring fully developed flow in open channels. Stream-flow technicians have to put significant effort into collecting stream discharge information at low flows to maintain a sufficient quality of data. Nonetheless, there is a great need for better low-flow estimation by many user groups of streamflow data, such as aquatic ecologists (Nilsson et al., 2003) and drought forecasters. There is also a need to collect information at sites other than gaging stations to develop low-flow estimation procedures.
- **Unit discharges.** In the current NWIS water information dissemination program, instantaneous discharges that provide essential hydrographic and peak flow information for streams are stored for 30 days after an event occurs. These data are essential for evaluation of channel stability, water and sediment routing, and so forth. **These data should be archived electronically in a retrievable form.** This is not as great an effort as it may seem. Daily mean discharge data are compiled by using the rating curve to convert each recorded stage value (e.g., each 15 minutes) to a corresponding discharge value. The resulting discharge values (the "unit values") are then averaged over a day to give the published value of daily mean

discharge. The quality control process of checking that the recorded stage values are valid and that the rating curve is appropriate is already being carried out at the level of unit values so no further quality assurance would be needed if these values were published rather than simply the daily mean discharge values.

- **Crest stage data.** Crest stage data have been collected by the USGS in both past and present times. These data should be electronically archived and disseminated with other streamflow information. For some ephemeral streams, even in large watersheds, these may be the only data available. Further, as information technology continues to expand, historical records of extreme events will become increasingly important to researchers.

In addition to collecting and reporting streamflow data, the USGS typically collects non-flow information at NSIP gages, but much of this is not disseminated, which stymies advances in River Science. Collected but unreported information includes data on channel cross sections collected at gaging stations, bed particle size information, and flood survey data.

Importance of the Non-base NSIP Network for River Science

In previous parts of this document, we evaluate the core, or base, NSIP network. It should be emphasized, however, that the non-base network is also essential to evaluate regional channel geometry relationships, downstream changes in surface water-groundwater interactions, and other river science relationships.

For example, Andrews et al. (2004) used 38 non-NSIP gages in California to examine the influence of the El Niño-Southern Oscillation (ENSO) phase on flooding in coastal streams. They created a normalized El Niño flood magnitude for various recurrence intervals as the ratio of twice the El Niño flood divided by the sum of the El Niño and non-El Niño floods. The relative magnitude of El Niño floods with a five-year recurrence interval decreases with latitude (Figure 6-6), which explains 84 percent of the variation in relative flood magnitude between El Niño and non-El Niño phases in California coastal streams. This analysis further showed that depending on local orographic effects, ENSO floods can be significantly smaller than expected solely from latitudinal position, as seen for floods in Soquel and Corralitos Creeks, which lie in a rain shadow.

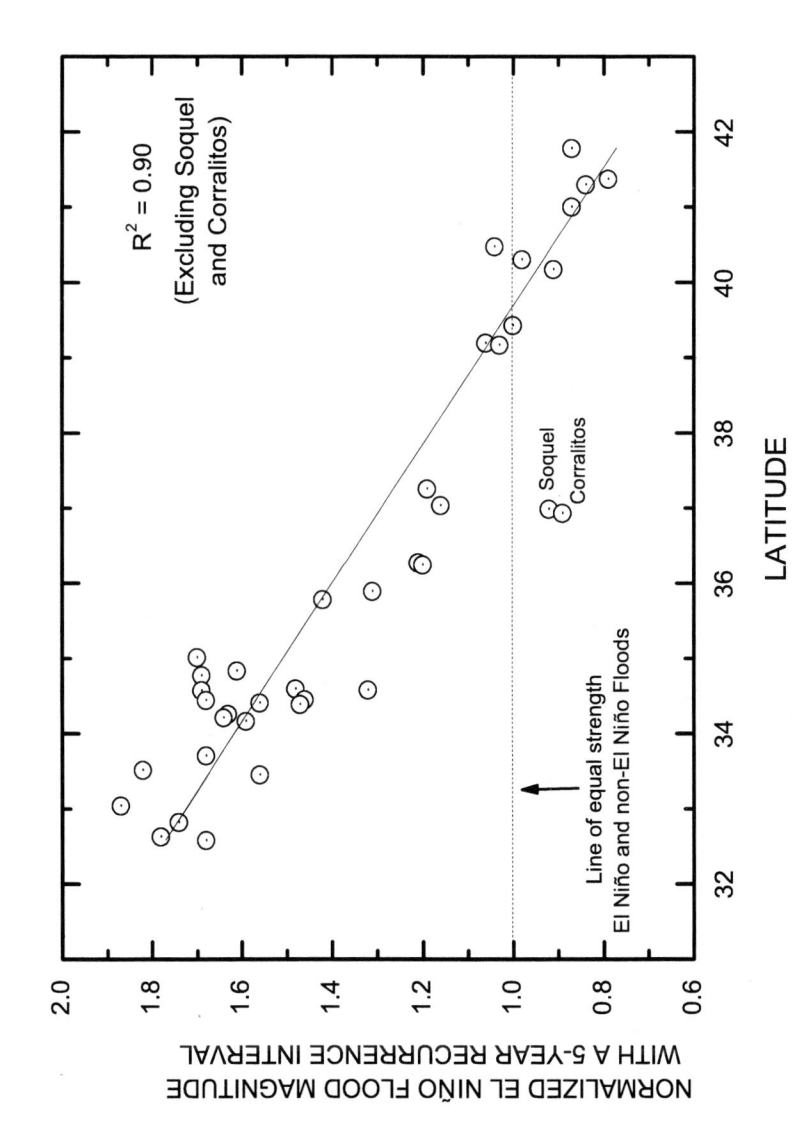

FIGURE 6-6 Relationship between normalized El Niño flood magnitude and latitude for 38 California coastal streams. SOURCE: Andrews et al. (in press).

SUMMARY

The NSIP data management system should be developed or designed with the capacity to integrate nontraditional or emerging data types, such as satellite imagery, velocity profiles from ADCM's, particle size information, channel mapping, etc. Developing and new technologies will require the capacity to store, manipulate, and disseminate more than simply "tabular records."

Data of relevance to river science that have not been archived electronically should be rescued, if necessary, by digitizing from paper records and made available on the Internet. Valuable information is contained in crest stage data, slope-area data from flood studies, and gaging station channel geometry and bed sediment characteristics.

The USGS should to continue to work on explicitly linking surface water to groundwater. This should be done in the context both of gages (estimating groundwater inputs) and of modeling.

The USGS should identify watersheds for which good hydrological information is available and land-use changes are documented. These sites should be prime sites at which hydrographic information is retrieved and stored to better understand how changes in land use affect hydrological characteristics. This will improve both planning and knowledge of the ecological and geomorphic consequences of land-use changes.

With the addition of channel morphology data, sentinel watersheds (Goal 4 of the NSIP) can provide not only hydrological reference points for the nation but stream morphology reference points as well. The representativeness of sentinel watersheds for characterizing the hydrologic and geomorphic diversity of the nation in support of river science should be explicitly evaluated.

Finally, this chapter raises as many questions as it answers. For example, which kinds of integrative river science questions should be investigated at the USGS and which are more appropriate for the broader scientific community? Within the USGS, how can monitoring efforts involving flow, sediment, chemistry, and biota be integrated? Also, what temporal and spatial scales should the USGS focus on? These are just three of a substantial set of issues that the USGS will have to resolve in order to design a truly effective program in the multidisciplinary science of rivers.

7
Summary and Conclusions

The National Streamflow Information Program (NSIP) has been formulated by the U.S. Geological Survey (USGS) to create a stable, federally funded *base network* of streamgages and to enhance the information derived from this network with intensive data collection during major floods and droughts, periodic regional and national assessments of streamflow characteristics, enhanced streamflow information delivery to customers, and methods development and research. The USGS has proposed that the base network of streamgages meet five minimum federal streamflow information goals, namely, (1) interstate and international agreements, (2) flow forecasts, (3) river basin outflows, (4) long-term monitoring using benchmark (sentinel) watersheds, and (5) water quality. This report examines the goals and method by which the base gage network sites were selected, the rationale for the supporting elements of the NSIP, and the role of streamflow information in advancing river science.

The USGS is the nation's unquestioned leader in the conduct of streamgaging, and its national repository of streamflow information has in recent years been made much more accessible to the public through an exemplary program of information publication on the Internet. Overall, the committee concludes that the National Streamflow Information Program is a sound, well-conceived program that meets the nation's needs for streamflow measurement, interpretation, and information delivery.

The nation needs streamflow information to address water management issues related to irrigation, flood warning, public water supply, water-power generation, water conservation, industrial water supply, chemical loading, recreation, and biological health of rivers. For more than a century, the USGS has met this task by developing and maintaining the national streamgaging program, and by publishing the resulting data. Also, USGS research has enhanced the understanding of river processes through the publication of thousands of documents on the state and behavior of the nation's waterways. This research rests on the foundation of a network of gages and a large body of water quality and river sediment data.

RATIONALE FOR FEDERAL SUPPORT OF A NATIONAL NETWORK

A strong federal role in the streamgaging network is important in view of the growing stress on water resources arising from population expansion and movement into water-short and flood-prone areas. In the words of one federal flood forecaster, "[USGS streamgages] are everything; without them we are dead in the water" (Gary McDevitt, National Weather Service [NWS] River Forecast Center, Chanhassen Minn., oral communication, 2002). Therefore, there should be national support for a base network of permanent gages. However, the USGS, in collaboration with the NWS, needs to communicate better that streamflow information creates public value, for example, by saving lives and preventing economic losses through flood forecasting.

A federal agency logically fills the role of providing streamflow information because such information supports national interests, not just local or private interests. In fact, streamflow information has many of the properties of a public good, because everyone benefits, whether they pay or not, and benefits to additional "users" come at no additional cost. The public also values efficiency and equality of access, both of which are characteristics of federally provided information. National interests are served by the provision of impartial, legally accepted information for arbitration of interstate water supply disputes. Streamflow information is also essential for state and local water supply management, and consequently many USGS gages are partly funded by local cooperators.

The streamgaging network, however, has had to contend with unstable and discontinuous funding support. Gages have been inactivated when cooperators cut budgets, and these incremental losses have eroded the network. Many inactivated gages had long records that are valuable for trend analysis and forecasting. It is practically impossible to quantify the cost of losing an individual gage. Its value even for one goal—for example, flood or drought forecasting—is embedded in the operation and accuracy of the entire forecast system, the forecast delivery mechanisms, and the forecast response. It is the integrity of the system as a whole that must be safeguarded.

Federal support of the base network would help provide stability and continuity to the network. Federal support of part of the network does not, however, imply the sufficiency of the overall network, which will always rely heavily on cooperators to help meet national goals for stream data.

THE BASE GAGE NETWORK

There are about 7,300 USGS-operated streamgages presently recording continuous stage and flow data. Not all of these would be federally funded base gages in the NSIP. Only 5,293 gage sites are listed under the five NSIP criteria, and since some sites serve more than one criterion, the actual number of sites presently identified as NSIP base gages is 4,424. Further, about 1,300 of these sites are not active: about 800 are inactive and 500 would be new. Of the remaining 3,100 or so currently active gages, the NSIP base gage network includes 2,800 gages that the USGS presently operates and 300 gages that other agencies operate and for which, under a fully funded NSIP, the USGS would assume the operational costs.

One concludes that the majority of the 7,300 USGS-operated streamgages will not form part of the base gage network. This does not mean that they are not fulfilling important purposes, but simply that those purposes may be primarily local in scale or otherwise not of highest national priority as defined by the five federal goals noted above. Regardless, all active USGS streamgages are considered to be part of the overall NSIP network.

In the following sections, each stated NSIP goal for the base gage network and the number of gage sites designated to meet that goal are examined in turn.

NSIP Goal 1: Meeting Legal and Treaty Obligations on Interstate and International Waters

The USGS designates 515 gage sites to provide streamflow information supporting legal compacts (185 gages) or to gage flow near where a stream crosses a state or international border if the upstream drainage area exceeds 500 square miles (330 gages). An examination of the NSIP base gage network was also conducted by the Interstate Council on Water Policy (ICWP, 2002). It concluded that there is not a compelling federal need for providing streamflow information at state and international borders with no legal compacts.

The committee does not concur with this view and believes that the USGS should proceed with the NSIP gage sites at their planned locations. Water-use permitting and data collection practices vary greatly from state to state. USGS streamflow data have been critical in cases of interstate disputes, especially during drought. As competition for water increases over time, further interstate conflict over water use will likely arise. Resolving such disputes will rest on a foundation of long-term streamflow information, and it will be too late once the conflict emerges to begin to collect such information.

NSIP Goal 2: Flow Forecasting

The USGS has designated 3,244 gage sites as part of the base network to support the flow forecasting mission of the National Weather Service. This number is 73 percent of the 4,424 NSIP base gage network sites, so it is clear that this goal dominates numerically among the five NSIP site selection goals. The USGS and the NWS have complementary roles with respect to streamflow information: the USGS does streamflow measurement and the NWS does flood forecasting. Thus, the USGS deals with past and present (real-time) streamflow information, and the NWS focuses on the near-term future. The NWS operates hydrologic models whose forecast points at watershed outlets are located wherever possible at USGS streamgage sites. As part of the flow forecasting goal, the USGS intends to provide streamgaging data at all NWS forecast points. The NWS hydrologic models also forecast flow "data points," which are the outlets of other watersheds used in the hydrologic model. Many of these points are also located at USGS gage sites so as to allow for forecast model calibration. Thus, USGS gage information is crucial to the NWS flood forecasting mission. With nationwide losses due to flooding averaging on the order of $1 billion per year in recent years, this goal is well justified as a criterion for NSIP gage selection.

The U.S. Department of Agriculture (USDA) National Resource Conservation Service (NRCS) also has a forecasting mission in the western states for estimating water supply over the coming months. This mission involves 576 forecast sites, of which 321 are already included in the NSIP base gage network. Since the NSIP mission is to support flow forecasting, as distinct from just flood forecasting, the NRCS forecast sites should also be included in the NSIP base gage network. This would add 255 new sites to the 3,244 sites presently attributed to the flow forecasting goal, an increase of 8 percent. A joint task force of the three agencies is needed to prioritize the addition of gages at the flow forecasting sites.

NSIP Goal 3: Measuring River Basin Outflows

The USGS designates 450 gages to measure discharge from major watersheds. Streamgaging sites are designated near the outflow of each of the nation's 352 Hydrologic Accounting Units (six-digit hydrologic unit code basins). Adequate coverage will allow the USGS to calculate regional water balances over the nation. Federally supported, long-term gages provide the continuity needed to calculate present and forecast future river basin outflows. River basin outflows over different time scales are the integrated response of the entire hydrologic system within the basins. Knowing how outflows are affected by changes in climate and landcover will lead to better forecasting and contribute to a better understanding of regional differences in hydrologic systems. Stream basins are inherently "nested," with large basins encompassing smaller ones. Knowing how and why outflows change per unit area from small-size to large-size basins will lead to better extrapolation of extreme floods and low flows. Overall, the breadth of ongoing and potential applications of a sound understanding of the hydrologic response of basins throughout the country justifies the inclusion of this goal as a selection criterion for the NSIP.

NSIP Goal 4: Monitoring Sentinel Watersheds

Sentinel watersheds are those watersheds chosen to represent the hydrologic diversity in the nation's landscape. The USGS designates 874 gage sites to meet this goal. The criteria for selecting sentinel watersheds are watershed size and representation of ecoregions. Watersheds with regulated (e.g., dammed) streams are avoided, and preference is given to watersheds that have been minimally influenced by human activities, thereby allowing tracking of long-term trends. Sentinel watersheds, which may also serve other roles, provide important information to meet long-term national needs for monitoring and science. In particular, long-term streamflow records in sentinel watersheds provide the benchmark data needed to assess hydrologic, ecologic, and water quality changes in similar, more numerous, watersheds with substantial anthropogenic landscape changes and thereby improve watershed management and planning. Given the interplay between hydrology and geomorphology, collecting channel morphological data in the sentinel watersheds would increase their scientific value—the sentinel watersheds could serve not only as hydrologic reference sites, but also as morphologic reference sites.

NSIP Goal 5: Measuring Flow for Water Quality

Water quality is closely tied to a stream's discharge, which dictates the concentration and flux of pollutants. High discharges may dilute pollutants; low discharges may concentrate them. On the other hand, pollutant loads (e.g., from agricultural or urban runoff) may increase under high-flow conditions. Proper interpretation of water quality data requires knowledge of stream discharge. The USGS designates 210 gage sites to provide streamflow information for a national network of water quality (concentration and loading) monitoring points. This streamflow information is matched to three national water quality networks: Hydrologic Benchmark (HBM) (63 stations), National Stream Water Quality Accounting Network (NASQAN) (40 stations), and National Water Quality Assessment Low-Intensity Phase (NAWQA-LIP) (107 stations).

The NSIP also supports other water-quality needs. For example, the Total Maximum Daily Load (TMDL) program of the Environmental Protection Agency (EPA) requires estimates of flow to determine chemical loads and transport. However, additional gaging to quantify the inflow to every one of the thousands of impaired water segments included in the TMDL program would be overwhelming in cost and manpower. There is a pressing need to be able to spatially interpolate streamflow time series from gaged locations to any point on the river network. Advances in geospatial information processing, used by the USGS in the NSIP site selection process, can be adapted for this purpose, as the USGS is doing in its Streamstats program for estimating streamflow statistics at ungaged locations. The USGS is well positioned in terms of expertise to do this research.

Distribution of Gage Site Locations

In general, the distribution of gages by state across the nation produced by the NSIP criteria appears reasonable when measured on metrics of number of gages per unit of land area and number of persons per gage. A possible exception is Nevada, where the committee's analysis of the NSIP base gage network found a surprisingly small number of gage sites (30) relative to neighboring states—Arizona (85), Utah (111), Idaho (95), and Oregon (136).

This anomaly arises in part because the NWS has only 10 forecast points in Nevada, compared to an average of 74 in the four neighboring states. It also arises because many of the border gages between Nevada and adjacent states are located in the adjacent state rather than in Nevada, and because Nevada has only a small number of ecological zones, so there are fewer sentinel watershed gages than would otherwise be the case. If NRCS forecast sites are added to the flow forecasting goal, this would add 17 sites in Nevada for a total of 47 NSIP sites, which is less anomalous.

Nevada is the nation's driest state, so the low number of NSIP sites may also arise because many of the state's streams are ephemeral. The hydrologic characteristics of ephemeral streams throughout much of the greater southwestern United States are sparsely measured, and the NSIP should incorporate a strategy to begin evaluating this large hydrologic landscape through the sentinel gage program. A single set of rules for siting NSIP base gage sites across the country may in some regions have to be adapted to allow for special hydrologic conditions not experienced everywhere.

Base Gage Network Design Methods

The five proposed NSIP goals in the design of a *national* streamflow information base network are sound. With the possible exception of Nevada, the geographic distribution of gages produced by these NSIP goals appears reasonable when states are compared using metrics such as number of gages per unit of land area or number of persons per gage.

The USGS has developed an innovative method for selecting sites for the NSIP base gage network using geospatial analysis of the national stream network, drainage areas, ecological zones, and gage sites where other functions are performed, such as forecasting floods or systematic collection of water quality data. Historically, gage networks have most often been analyzed statistically, so the move to a geospatial analysis of gage sites is a significant departure from past practice in this field, but one that is in harmony with the advancement of geospatial information availability and analysis capabilities. There is a duality between the selection of sites in a network, and the delineation of subwatersheds draining to those sites, that defines the *coverage* of the NSIP base gage network.

Coverage models have been used in other site selection processes, such as the locations of fire stations within a city, where each fire station is associated with its service area. An advantage of the coverage approach to streamgage network design is that it identifies where gages should be located, rather than being limited to consideration of where they are located now. By creating national NSIP subwatershed dataset maps for each criterion using the proposed and active gage sites, the USGS can assess the completeness of coverage. When new gages are to be installed from the

NSIP site set, consideration can be given to the impact of site choice on the NSIP subwatershed dataset.

Statistical methods for stream network design are useful for ranking gages in order of their regional information content, as illustrated in this report by review of a statistically based streamgage network analysis for Texas. Statistical rankings help to identify which inactive gage sites should be activated first when additional funds to support NSIP gages become available, with the goal of maximizing the value of streamflow information while minimizing cost.

A new research initiative to regionalize streamflow characteristics is recommended, with the goal of being able to estimate streamflow time series and stream channel characteristics at any location on a stream or river in the United States with a quantitative estimate of uncertainty. Regionalization methods will significantly increase streamflow information coverage of the nation.

OTHER NSIP COMPONENTS

Besides enhancing the base gage network, the NSIP has four other components dealing with intensive data collection during major floods and droughts, assessments of streamflow characteristics, streamflow information delivery to customers, and methods development and research. These appropriate activities continue the USGS tradition of striving to improve the coverage, access, and quality of streamflow information.

In general, the strong efforts that the USGS has made to transform the National Streamflow Information Program from a "streamgaging program" to an integrated effort in which *information* products of various kinds are available when and where the user wants them are commendable. Likewise, the USGS's ongoing development of new ways of employing advanced technology to improve measurement and information delivery deserve credit.

The spatial scale and risks of hydrologic extremes (e.g., floods and droughts) are research areas deserving of the attention that the USGS proposes in the NSIP. The hydrologic system organizes itself spatially and dynamically such that the most extreme events are organized over the largest spatial and temporal scales. This task recognizes that the regional information content of the network is greater than the sum of the information from individual stations.

The USGS should further refine its information delivery strategy. If the NSIP goal is saving life and property as well as promoting prosperity and well-being, delivery of information is at least as important as data analysis.

This would include on-line, value-added products such as flood simulations and water supply and water quality projections under various development scenarios.

The USGS should disseminate more types of data, including historical data (requiring rescue of older paper format data), cross sections, velocity profiles, unit discharge values, and opportunistic data (e.g., crest stage data and slope-area data from flood studies). These data are essential to document channel changes, evaluate stream hydrographs, calculate hydraulic parameters, examine climate change, and infer certain hydroecological relationships. An NSIP data management system must be developed to accommodate various types and formats of data that support river science. A system for publishing the unit value data will allow users to obtain historical streamflow data for intervals of less than one day.

Streamgages are nodes in the streamgaging network, so the accuracy of information provided by the network rests on the quality and type of information provided by the gages themselves. Gages are traditionally viewed as stationary points gathering data in a method similar to that of 150 years ago. The USGS is attempting to develop the "gaging station of the future." There are many research opportunities for advancement over current methods:

- Develop and use a portfolio of data collection tools in addition to the fixed, permanent stations. This would include phasing in new technologies such as acoustic Doppler current profilers to measure stream velocity and channel resistance; making the data widely available to foster research outside USGS on the relationships among channel morphology, velocity, and flow resistance in channels; and providing real-time information delivery at critical stations through satellite links.

- Provide real-time water quality estimates analogous to those for streamflow. This is a very valuable adjunct to traditional streamflow information and, to the extent that resources permit, this capability should be expanded to other gages. Gages in areas prone to flash flooding should be equipped with critical-stage alarms or web cameras to alert the public and resource managers of impending hazards.

- Measure streamflow at ungaged sites during high- and low-flow conditions using mobile units to respond to events as they occur. These additional data also will assist in regionalizing streamflow characteristics.

The NSIP program will lead to advancements in all of these areas, and if due care is taken to ensure comparability between traditional streamgaging data and those of new technologies, these areas of research should be pursued.

ADAPTIVE MANAGEMENT

Overall the five components of the NSIP plan are well conceived and form strongly complementary program elements. Active integrated management and coordination will result in an information program that will generate value to the nation far greater than the sum of its parts. Nevertheless, information needs and technologies evolve rapidly and dynamically, and will continue to do so. This requires continuous improvement and coordination to maximize the value of the national investment in streamflow information. No single solution will meet all of the nation's needs for streamflow information or remain the best choice in the face of changing demands. The combination of dynamically changing demands with future uncertainty strongly suggests the need to develop, integrate, and use formal adaptive management techniques as an integral part of the NSIP. Adaptive management not only identifies goals and program components (as does the NSIP plan), but also identifies expected outcomes that can be described with meaningful performance measures. These provide a benchmark against which management decisions may be consistently revisited and re-evaluated relative to a more stable and clearly articulated set of goals and expected outcomes.

For example, one way to site gages is to identify point locations at which streamflow information would be useful—locating one continuous streamgage at each such point. The ICWP thereby identified the need for more than 18,000 gages. However, some of these information needs (e.g., for National Flood Insurance Program communities or Impaired Water Quality Reaches) can be satisfied (with some difference in the quality of information) with other techniques such as regionalization. The overarching goal for the NSIP should be to provide streamflow information (with quantitative confidence limits) at any arbitrary point on the landscape. The streamgage network must be sufficient to support this goal.

Adaptive management would identify the information need, determine the mode of information generation and delivery (e.g. gaging, spot measurements, indirect methods, hydrologic estimation) in order to achieve performance criteria, and later evaluate the expected and actual performance to determine whether modification is needed. It would help balance the multiple attributes of information—quality, reproducibility, resilience to extremes, and cost objective—and align resources to outcomes (not just activities). Implementation of adaptive management will generate performance information about the NSIP that will be essential to evaluate and incrementally improve the program in the future.

In addition, the USGS should consider how the public, the scientific community, and water management agencies will be included in the adap-

Summary and Conclusions

tive management of this national network. At present, much of the public input on prioritizing streamflow gaging comes in the form of having paying state and local customers through the Cooperative Water (Coop) Program. If the NSIP fully funds its base network independent of cost matching, other mechanisms for public consultation at various levels (e.g., an advisory board, surveys) will have to be found.

In summary, adaptive management and periodic systematic reevaluation should be an integral part of the program from its inception.

RIVER SCIENCE

The USGS has a long history of research on rivers. Pressing issues such as streamflow losses to groundwater pumping, nonpoint source pollution loads, and aquatic and riparian ecosystem degradation make a compelling case for developing river science. Streamflow information is a critical component supporting river science.

Streamflow information should be collected to promote an integrated, process-based understanding of hydrologic-geomorphic-biological linkages. Stream gradient, bed material size, and sediment transport should all be measured at more locations where discharge and stage are measured. Such data are needed for sediment and hydrologic routing models. The temporal and spatial characteristics of this material routing are of central importance to understanding many key ecological processes that influence ecosystem resilience and provide ecosystem goods and services.

Theoretical and empirical models are needed to estimate streamflow and channel characteristics *at any location* on the principal streams or rivers of the nation. Process-based models extend the value of streamflow data and support the generation of streamflow information throughout the watershed system.

To determine what data are most valuable, the USGS should engage the broader scientific community to seek input into what data it should be collecting for the development of river science. Since groundwater and surface water are two components of a fully integrated hydrologic system, appropriate data should be collected to understand aquifer-stream interactions.

In order to improve planning and assess the ecological and geomorphic consequences of land-use changes, the USGS should identify watersheds for which good hydrologic information is available and where land-use changes are documented. This information will improve understanding of how changes in land use affect hydrologic characteristics.

References

Adrian, R. J. 1984. Scattering particle characteristics and their effect on pulsed laser measurement of fluid flow: speckle velocimetry vs. particle image velocimetry. Applied Optics 23(11):1690-1691.

Alidi, S. 1993. Locating oil-spill response centers using mathematical models. Marine Pollution Bulletin 26(4):216-219.

Alsdorf, D. 2003. Water storage of the central Amazon floodplain measured with GIS and remote sensing imagery. Annals of the Association of American Geographers 93:55-66.

Alsdorf, D. E., J. M. Melack, T. Dunne, L. A. K. Mertes, L. L. Hess, and L. C. Smith. 2000. Interferometric radar measurements of water level changes on the Amazon flood plain. Nature 404:174-177.

Andrews, E. D., R. C. Antweiler, P. J. Neiman, and F. M. Ralph. 2004. Influence of ENSO on flood frequency along the California Coast. Journal of Climate 17(2): 337-348.

Auble, G. T., and M. L. Scott. 1998. Fluvial disturbance patches and cottonwood recruitment along the upper Missouri River, Montana. Wetlands 18(4):546-556.

Baron J. S., L. N. Poff, P. L. Angermeier, C. N. Dahm, P. H. Gleick, N. G. Hairston, Jr., R. B. Jackson, C. A. Johnston, B. G. Richter, and A. D. Steinman. 2002. Meeting ecological and societal needs for freshwater. Ecol. Appl. 12:1247-1260.

Bastin, G., B. Lorent, C. Duque, and M. Gevers. 1984. Optimal estimation of the average rainfall and optimal selection of raingauge locations. Water Resources Research 20(4):463-470.

Battin, T. J. 1999. Hydrologic flow paths control dissolved organic carbon fluxes and metabolism in an alpine stream hyporheic zone. Water Resources Research 35:3159-3169.

Berman, O., M. J. Hodgson, and D. Krass. 1995. Flow-interception problems. Pp. 389-426 in Facility Location: A Survey of Applications and Methods. New York: Springer.

Blasch, K. W., T. P. A. Ferré, A. H. Christensen, and J. P. Hoffmann. 2002. New field method to determine streamflow timing using electrical resistance sensors. Vadose Zone Journal 1:289-299.

Blasi, C. 2002. Personal Communication. Chairman of the LAWA Joint Water Commission of the Federal States-Committee for developing "Criteria Catalogue of Gauging Stations in Coastal Areas".

Boer, E. P. J., A. L. Dekkers, and A. Stein. 2002. Optimization of a monitoring network for sulfur dioxide. Journal of Environmental Quality 31(1):121-128.

Bosch, D. J. 1991. Benefits of transferring streamflow priority from agriculture to nonagricultural use. Water Resources Bulletin 27(3):397-405.

Bovee, K. D., and M. L. Scott. 2002. Implications of flood pulse restoration for populus regeneration on the Upper Missouri River. River Research and Applications 18:287-298.

Bradley, A. A. 1998. Regional frequency analysis methods for evaluating changes in hydrologic extremes. Water Resources Research 35(4):741-750.

Bradley, A. A., A. Kruger, E. A. Meselhe, and M. V. I. Muste. 2002. Flow measurement in streams using video imagery. Water Resources Research 38(12):1315.

Branas, C. C., E. J. MacKenzie, and C. S. Revelle. 2000. A trauma resource allocation model for ambulances and hospitals. Health Services Research 35(2):489-507.

Bras, R. F., and I. Rodriguez-Iturbe. 1976. Network design for the estimation of areal mean rainfall events. Water Resources Research 12(6):1185-1195.

Buchanan, T. J., and W. P. Somers. 1969. Discharge measurements at gauging stations. USGS Techniques of Water Resources Investigations, Chapter A8, Book 3. Reston, VA: USGS.

Bueso, M. C., J. M. Angulo, and F. J. Alonso. 1998. A state-space model approach to optimum spatial sampling design based on entropy. Environmental and Ecological Statistics 5(1):29-44.

Bunn, S. E., and A. H. Arthington. 2002. Basic principles and ecological consequences of altered flow regimes for aquatic biodiversity. Environ. Manage. 30:492-507.

Campbell, J. F., A. T. Ernst, and M. Krishnamoorthy. 2002. Hub location problems. Pp. 373-408 in Facility Location: Applications and Theory, Z. Drezner and H.W. Hamacher (eds.). Berlin, New York: Springer.

Cayan, D. R., S. A. Kammerdiener, M. D. Dettinger, J. M. Caprio, and D. H. Peterson. 2001. Changes in the onset of spring in the western United States. Bulletin of the American Meteorological Society 82(3):399-415.

Chiew, F., and T. McMahon. 1996. Trends in historical streamflow records. Pp. 63-69 in Regional Hydrological Response to Climate Change, J. Jones, C. Liu, M. K. Woo, and H. T. Kung (eds.). Dordrecht: Kluwer Academic.

Christensen, V. G., X. Jian, and A. C. Ziegler. 2000. Regression analysis and real-time water-quality monitoring to estimate constituent concentrations, loads, and yields in the Little Arkansas River, South-Central Kansas, 1995-1999. Water Resources Investigations Report 00-4126. Lawrence, KS: U.S. Geological Survey. 36 pp.

Christensen, V. G., P. P. Rasmussen, and A. C. Ziegler. 2002. Real-time water quality monitoring and regression analysis to estimate nutrient and bacteria concentrations in Kansas streams. Water Science & Technology 45(9):205–219.

Cirmo, C. P., and J. J. McDonald. 1997. Linking the hydrologic and biogeochemical controls of nitrogen transport in near-stream zones of temperate-forested catchments: a review. Journal of Hydrology 199:88-120.

Clausen, B., and B. J. F. Biggs. 2000. Flow indices for ecological studies in temperate streams: groupings based on covariance. Journal of Hydrology 237:184-197.

Cleaves, E. T., and E. J. Doheny. 2000. A Strategy for a Stream-Gaging Network in Maryland. Report of Investigations 71. Baltimore: Maryland Geological Survey. *(http://www.mgs.md.gov/esic/publications/new/ri71sum.html)*.

Cleveland, T. G., and W. Yeh. 1990. Sampling network design for transport parameter identification. Journal of Water Resources Planning and Management 116(6):765-783.

Cloke, P. S., and I. Cordery. 1993. The value of streamflow data for storage design. Water Resources Research 29(7):2371-2376.

Constantz, J., D. Stonestrom, A. E. Stewart, R. Niswonger, and T. R. Smith. 2001. Analysis of streambed temperature in ephemeral stream channels to determine streamflow frequency and duration. Water Resources Research 37:317–328.

Cordery, I., and P. S. Cloke. 1992. Economics of streamflow data-collection. Water International 17(1):28-32.

Costa, J. E., K. R. Spicer, R. T. Cheng, F. P. Haeni, N. B. Melcher, E. M. Thurman, W. J. Plant, and W. C. Keller. 2000. Measuring stream discharge by non-contact methods: a proof-of-concept experiment. Geophysical Research Letters 27(4):553-556.

Cover, T., and J. Thomas. 1991. Elements of Information Theory. New York: John Wiley & Sons.
Creutin, J. D., M. V. Muste, A. A. Bradley, S. C. Kim, and A. Kruger. 2003. River gauging using PIV technique: proof of concept experiment on the Iowa River. Journal of Hydrology 277(3-4):182-194.
Current, J., and M. O'Kelly. Locating emergency warning sirens. 1992. Decision Sciences 23(1):221-234.
Daskin, M. S. 1983. A maximum expected covering location model—formulation, properties and heuristic solution. Transport Science 17(1):48-70.
Dettinger, M. D., and D. R. Cayan. 1995. Large-scale atmospheric forcing of recent trends toward early snowmelt runoff in California. Journal of Climate 8(3):606-623.
Doheny, E. J. 1998. Evaluation of the stream-gauging network in Delaware. Report of Investigations No. 57. Newark, NJ: Delaware Geological Survey. 32 pp.
Douglas, A. J., and J. G. Taylor. 1998. Riverine based eco-tourism: Trinity River non-market benefits estimates. International Journal of Sustainable Development and World Ecology 5(2):136-148.
Drabek, T. E. 1999. Understanding disaster warning response. Social Science Journal 36(3):515-523.
Drezner, Z. (ed.). 1995. Facility location. Berlin: Springer Verlag.
Drezner, Z., and H. W. Hamacher (eds.). 2001. Facility Location: Theory and Algorithms. Berlin: Springer Verlag. 547 pp.
Duff, J. H., F. Murphy, C. C. Fuller, F. J. Triska, J. W. Harvey, and A. P. Jackman. 1998. A mini drivepoint sampler for measuring pore water solute concentrations in the hyporheic zone of sand-bottom streams. Limnology and Oceanography 43(6):1378-1383.
Eash, D. A. 1997. Effects of the 1993 Flood on the Determination of Flood Magnitude and Frequency in Iowa—Floods in the Upper Mississippi River Basin 1993. USGS Circular 1120-k. Reston, VA: U.S. Geological Survey.
Federal Emergency Management Agency (FEMA). 2003. http://www.fema.gov/nfip.
Fontaine, R. A., M. E. Moss, J. A. Smith, W.O. Thomas, Jr. 1984. Cost Effectiveness of the Stream-Gaging Program in Maine. USGS Water Supply Paper 2244. Reston, VA: U.S. Geological Survey.
Friedman, J. M., and G. T. Auble. 1999. Mortality of riparian trees by sediment mobilization and extended inundation. Regulated Rivers: Research and Management 15:463-476.
Gasith A., and V. H. Resh. 1999. Streams in Mediterranean climate re-

gions: abiotic influences and biotic responses to predictable seasonal events. Annual Review of Ecology and Systematics 30:51-81.

Geist, D. R., M. C. Joy, D. R. Lee, and T. Gonser. 1998. A method for installing piezometers in large cobble bed rivers. Ground Water Monitoring & Remediation 18:78-82.

Gendreau M., G. Laporte, and I. Parent. 2000. Heuristics for the location of inspection stations on a network. Naval Research Logistics 47(4):287-303.

Gerhard, H. 2002. Personal Communication. Hessian Agency for the Environment and Geology. Director of the Division for Water, Waste, Contaminated Sites. Germany.

Ghosh A., and S. L. Mclafferty. 1982. Locating stores in uncertain environments—a scenario planning approach. Journal of Retailing 58(4):5-22.

Goolsby, D. A., and W. A. Battaglin. 2000. Nitrogen in the Mississippi Basin—Estimating Sources and Predicting Flux to the Gulf of Mexico, USGS Fact Sheet FS-135-00. Reston, VA: U.S. Geological Survey.

Goolsby, D. A., W. A. Battaglin, and E. M. Thurman. 1993. Occurrence and Transport of Agricultural Chemicals in the Mississippi River Basin July through August 1993. USGS Circular 1120-c. Reston, VA: USGS.

Grimm, N. B., and S. G. Fisher. 1984. Exchange between interstitial and surface waters— implications for stream metabolism and nutrient cycling. Hydrobiologia 111:219-228.

Gruntfest, E., and J. W. Handmer (eds.). 2001. Coping with Flash Floods. Dordrecht: Kluwer Academic Publishers.

Haas, T. C. 1992. Redesigning continental-scale monitoring networks. Atmospheric Environment Part A-General Topics 26(18):3323-3333.

Haeni, F. P., M. L. Buursink, J. L. Costa, N. B. Melcher, R. T. Cheng, and W. J. Plant. 2000. Ground-penetrating RADAR methods used in surface-water discharge measurements. Pp. 494-500 in Proceedings of the Eighth International Conference on Ground Penetrating Radar, D. A. Noon, G. F. Stickley, and D. Longstaff (eds.), Queensland, Australia: University of Queensland.

Haghani, A. 1996. Capacitated maximum covering location models: formulations and solution procedures. Journal of Advanced Transportation 30(3):101-136.

Hamlet, A. F., and D. P. Lettenmaier. 1999. Columbia River streamflow forecasting based on ENSO and PDO climate signals. J. Water Res. Plann. Mgmt. 125:333-341.

Hamlet, A. F., D. Huppert, and D. P. Lettenmaier. 2002. Economic value of long-lead streamflow forecasts for Columbia River hydropower.

Journal of Water Resources Planning and Management-ASCE 128(2): 91-101.

Handmer, J., C. Keys, and J. Elliott. 1999. Achieving lasting change in multi-organizational tasks: the case of flood warnings in Australia. Applied Geography 19(3):179-197.

Hansen, L. T., and A. Hallam. 1991. National estimates of the recreational value of streamflow. Water Resources Research 27(2):167-175.

Hardison, C. H., and M. E. Moss. 1972. Accuracy of low-flow characteristics estimated by correlation of base-flow measurements. USGS Water-Supply Paper 1542-B. Reston, VA: U.S. Geological Survey. 21 pp.

Harvey, J. W., and C. C. Fuller. 1998. Effect of enhanced manganese oxidation in the hyporheic zone on basin-scale geochemical mass balance. Water Resources Research 34: 623-636.

Havskov, J., L. B. Kvamme, R. A. Hansen, H. Bungum, and C. D. Lindholm. 1992. The northern Norway seismic network—design, operation, and results. Bulletin of the Seismological Society of America 82(1):481-496.

Hendricks, S. P. 1993. Microbial ecology of the hyporheic zone: a perspective integrating hydrology and biology. Journal of the North American Benthological Society 12:70-78.

Hill, A. R., C. F. Labadia, and K. Sanmugadas. 1998. Hyporheic zone hydrology and nitrogen dynamics in relation to the streambed topography of a N-rich stream. Biogeochemistry 42:285-310.

Hinkle, S. R., J. H. Duff, F. J. Triska, A. Laenen, E. B. Gates, K. E. Bencala, D. A. Wentz, and S. R. Silva. 2001. Linking hyporheic flow and nitrogen cycling near the Willamette River: a large river in Oregon, USA. Journal of Hydrology 244(3-4):157-180.

Hodgson M. J., K. E. Rosing, and J. J. Zhang. 1996. Locating vehicle inspection stations to protect a transportation network. Geographical Analysis 28(4):299-314.

Hogan, K., and C. S. Revelle. 1986. Concepts and applications of backup coverage. Management Science 32(11):1434-1444.

Holmes, R. R., Jr. 1996. Sediment transport in the Lower Missouri and the Central Mississippi Rivers June 26 through September 14 1993. USGS Circular 1120-i. Reston, VA: U.S. Geological Survey.

Hren, J., T. H. Chaney, J. M. Norris, and C. J. O. Childress. 1987. Inventory and Evaluation of 1984 Programs and Costs, Phase I of Water-Quality Data-Collection activities in Colorado and Ohio. USGS Water-Supply Paper 2295--A, 71 p. Reston, VA: U.S. Geological Survey.

Hughes J. M. R., and B. James. 1989. A hydrological regionalization of

streams in Victoria, Australia with implication for stream ecology. Australian Journal of Marine and Freshwater Research 40:303-326.
Interstate Council on Water Policy (ICWP). 2002. A Critique of the USGS National Streamflow Information Program and Considerations in Establishing a National Streamgaging Network. Report to Federal Advisory Committee on Water Information. Washington, D.C.: Interstate Council on Water Policy. 13 pp.
Instream Flow Council (IFC). 2002. Instream flows for riverine resource stewardship. Available at http://www.instreamflowcouncil.org.
Jacobson, R. B., and K. A. Oberg. 1997. Geomorphic Changes on the Mississippi River Flood Plain at Miller City, Illinois as a Result of the Flood of 1993. USGS Circular 1120-j. Reston, VA: U.S. Geological Survey.
Jones, J. B., and P. J. Mulholland (eds.). 2000. Streams and Ground Waters. San Diego, CA: Academic Press. 425 pp.
Jones, J. L., J. M. Fulford, and F. D. Voss. 2002. Near-real-time simulation and Internet-based delivery of forecast-flood inundation maps using two-dimensional hydraulic modeling: a pilot study of the Snoqualmie River, Washington. Water Resources Investigations Report 02-4251. Tacoma, WA: U.S. Geological Survey. 35 pp.
Junk, W. J., P. B. Bayley, and R. E. Sparks. 1989. The flood pulse concept in river-floodplain systems. Canadian Special Publication of Fisheries and Aquatic Sciences 106:110-127.
Kahya, E., and J. A. Dracup. 1993. United States streamflow patterns in relation to the El Nino Southern Oscillation. Water Resources Research 29(8):2491-2503.
Karl, T. R., and R. W. Knight. 1998. Secular trend of precipitation amount, frequency, and intensity in the United States. Bulletin of the American Meteorological Society 79:231-242.
Kleindorfer, P. R., and H.C. Kunreuther. 1994. Siting of hazardous facilities. Pp. 403-440 in Operations Research and the Public Sector, S. M. Pollock, M. H. Rothkopf, and A. Barnett (eds.). Amsterdam, New York: North Holland.
Knapp, H. V., and M. Markus. 2003. Evaluation of the Illinois Streamflow Gaging Network. ISWS CR 2003-05. Champaign, IL: Illinois State Water Survey. 97 pp. Available at http://www.sws.uiuc.edu/pubdoc/CR/ISWSCR2003-05.pdf.
Kolm, S-Ch. 1988. Public economics. The New Palgrave Dictionary of Economics, vol 3, pp. 1049-1051, J. Eatwell, M. Milgate, and P. Newman (eds.), London: McMillan.
Kolpin, D. W., and E. M. Thurman. 1995. Postflood occurence of selected agricultural chemicals and volatile organic compounds in near-surface

unconsolidated aquifers in the Upper Mississippi River Basin 1993. USGS Circular 1120-g. Reston, VA: U.S. Geological Survey.

Krzysztofowicz, R. 1999. Bayesian theory of probabilistic forecasting via deterministic hydrologic model. Water Resources Research 35(9):2739-2750.

Kullback, S., and R. A. Leibler. 1951. On information and sufficiency. Annals of Mathematical Statistics 22:79-86.

Kunkel, K. E. 2003. North American trends in extreme precipitation. Natural Hazards 29:291-305.

Lee, J. 1998. Constrained maximum-entropy sampling. Operations Research 46(5):655-664.

Leones, J., B. Colby, D. Cory, and L. Ryan. 1997. Measuring regional economic impacts of streamflow depletions. Water Resources Research 33(4): 831-838.

Leopold, L.B., and T. Maddock. 1953. The hydraulic geometry of stream channels and some physiographic implications. U.S. Geological Survey Professional Paper No. 252. Washington, D.C.: U.S. Government Printing Office.

Lettenmaier, D. P., E. F. Wood, and J. R. Wallis. 1994. Hydro-climatological trends in the continental United States, 1948-1988. J. Clim. 7:586-607.

Lins, H. F. 1997. Regional streamflow regimes and hydroclimatology of the United States. Water Resources Research 33(7):1655-1667.

Lins, H. F., and P. J. Michaels. 1994. Increasing U.S. streamflow linked to greenhouse forcing. Eos Trans. American Geophysical Union 75:281, 284-285.

Lins, H. F., and J. R. Slack. 1999. Streamflow trends in the United States. Geophysical Research Letters 26(2):227-230.

Maidment, D. R. (ed.). 2002. Arc Hydro: GIS for Water Resources. Redlands, CA: ESRI Press.

Marianov, V., and C. S. Revelle. 1991. The standard response fire protection siting problem. Infor. 29(2):116-129.

Mason, R. R., Jr., and B. A. Weiger, 1995. Stream Gaging and Flood Forecasting. Fact Sheet FS 209-95. Reston, VA: U.S. Geological Survey and National Weather Service.

McCabe, G. J., and D. M. Wolock. 2002. A step increase in streamflow in the conterminous United States. Geophysical Research Letters 29 (24):2185.

Medina, K. D. 1987. Analysis of Surface-Water Data Network in Kansas for Effectiveness in Providing Regional Streamflow Information [with a section on Theory and Application of Generalized Least Squares by G. D. Tasker]. USGS Water-Supply Paper 2303. Reston, VA: U.S. Geological Survey. 28 pp.

Melcher, N. B., R. T. Cheng, and F. P. Haeni. 1999. Investigating technologies to monitor open-channel discharge by direct measurement of cross-sectional area and velocity of flow. In Hydraulic Engineering for Sustainable Water Resources Management at the Turn of the Millennium. Graz, Austria: Technical University Graz, Institute for Hydraulic and Hydrology.

Melcher, N. B., J. E. Costa, F. P. Haeni, R. T. Cheng, E. M. Thurman, M. Buursink, K. R. Spicer, E. Hayes, W. J. Plant, W. C. Keller, and K. Hayes. 2002. River Discharge Measurement with Helicopter-Mounted Radars. Unpublished manuscript. 15 pp.

Mertes, L. A. K., M. O. Smith, and J. B. Adams. 1993. Estimating suspended sediment concentrations in surface waters of the Amazon River wetlands from Landsat images. Remote Sensing of Environment 43(3)(199303):281-301.

Mirchandani, P. B., R. Rebello, and A. Agnetis. 1995. The inspection station location problem in hazardous material transportation—some heuristics and bounds. Infor. 33(2):100-113.

Mogheir, Y., and V. P. Singh. 2002. Application of information theory to groundwater quality monitoring networks. Water Resources Management 16:37-49.

Moody, J. A. 1995. Propagation and composition of the flood wave on the Upper Mississippi River 1993. USGS Circular 1120-f. Reston, VA: U.S. Geological Survey.

Moss, M. E. 1982. Concepts and techniques in hydrological network design. Operational Hydrology Report No. 19, WMO Report No. 580. Geneva, Switzerland: World Meteorological Organization. 30 p.

Mulvey J. M., R. J. Vanderbei, and S. A. Zenios. 1995. Robust optimization of large-scale systems. Operations Research 43(2):264-281.

Nagorski, S. A., and J. N. Moore. 1999. Arsenic mobilization in the hyporheic zone of a contaminated stream. Water Resources Research 35:3441-3450.

Naiman, R. J., S. E. Bunn, and C. Nilsson. 2002. Legitimizing fluvial ecosystems as users of water. Environmental Management 30:455-467.

National Research Council (NRC). 1992. Regional Hydrology and the USGS Stream Gaging Network. Washington, D.C.: National Academy Press.

National Research Council (NRC). 2001. Assessing the TMDL Approach to Water Quality Management. Washington, D.C.: National Academy Press.

National Research Council (NRC). 2002. Estimating Water Use in the United States: A New Paradigm for the National Water-Use Information Program. Washington, D.C.: National Academy Press.

National Weather Service (NWS). 2003. Service Hydrologist Reference Manual. NWS Office of Hydrology (http://www.nws.noaa.gov/oh/-hod/).

Nilsson C., J. E. Pizzuto, and G. Moglen. 2003. Ecological forecasting and running water systems: Challenges for economists, hydrologists, geomorphologists, and ecologists. Ecosystems.

Nixon, S.C. (ed.) 1999. European Freshwater monitoring Network Design. Topic report No. 10/1996. Copenhagen: European Environmental Agency.

O'Connor, J. E. and J. E. Costa. 2004. Spatial distribution of the largest rainfall-runoff floods from basins between 2.6 and 26,000 km^2 in the United States and Puerto Rico. Water Resources Research 40, W01107, doi:10.1029/2003WR002247.

Oehlert, G.W. 1996. Shrinking a wet deposition network. Atmospheric Environment 30(8):1347-1357.

Olden, J. D., and N. L. Poff. 2003. Redundancy and the choice of hydrologic indices for characterizing streamflow regimes. River Research and Applications 19:101-121.

Omernik, J. M. 1987. Aquatic ecoregions of the conterminous United States. Annals of the Association of American Geographers 77:118-125.

Owen, S. H., and M. S. Daskin. 1998. Strategic facility location: a review. European Journal of Operational Research 111(3):423-447.

Pardo-Igúzquiza, E. 1998. Optimal selection of number and location of rainfall gauges for areal rainfall estimation using geostatistics and simulated annealing. Journal of Hydrology 210(1998):206-220.

Parrett, C., N. B. Melcher, and R. W. James, Jr. 1993. Flood discharges in the Upper Mississippi River Basin—1993. USGS Circular 1120-a. Reston, VA: U.S. Geological Survey.

Pelletier, P. M. 1988. Uncertainties in the single determination of river discharge: a literature review. Canadian Journal of Civil Engineering 15(5):834-850.

Perez-Abreu, V., and J. E. Rodriguez. 1996. Index of effectiveness of a multivariate environmental monitoring network. Environmetrics 7(5): 489-501.

Perry, C. A. 1994. Effects of reservoirs on flood discharges in the Kansas and the Missouri River Basins. USGS Circular 1120-e. Reston, VA: U.S. Geological Survey.

Plant, W. J. 1990. Bragg scattering of electromagnetic waves from the air/sea interface. Pp. 41-108 in Surface Waves and Fluxes: Current Theory and Remote Sensing, G. L. Geernaert and W. J. Plant (eds.). Dordrecht: Kluwer Academic Publishers.

Poff, N. L. 1996. A hydrogeography of unregulated streams in the United States and an examination of scale-dependence in some hydrological descriptors. Freshwater Biology 36:71-91.

Poff, N. L., and J. V. Ward. 1989. Implications of streamflow variability and predictability for lotic community structure: a regional analysis of streamflow patterns. Canadian Journal of Fisheries and Aquatic Sciences 46:1805-1818.

Poff, N. L., and J. V. Ward. 1990. The physical habitat template of lotic systems: recovery in the context of historical pattern of spatio-temporal heterogeneity. Environmental Management 14:629-646.

Poff, N.L., J. D. Allan, and M. B. Bain. 1997. The natural flow regime: a paradigm for river conservation and restoration. BioScience 47:769-784.

Poff, N. L., J. D. Allan, M. A. Palmer, D. D. Hart, B. D. Richter, A. H. Arthington, J. L. Meyer, J. A. Stanford, and K. H. Rogers. 2003. River flows and water wars: emerging science for environmental decision-making. Frontiers in Ecology and the Environment 1: 298-306.

Postel, S. L., G. C. Daily, and P. R. Ehrlich. 1996. Human appropriation of available fresh water. Science 271:785-788.

Potter, K. W. 2001. A simple method for estimating baseflow at ungaged locations. Journal of the American Water Resources Association 37(1):177-184.

Rabbitt, M. C. 1989. The United States Geological Survey: 1879-1989. USGS Circular 1050. Reston, VA: U.S. Geological Survey.

Redmond, K. T., and K. W. Koch. 1991. Surface climate and streamflow variability in the western United States and their relationship to large-scale circulation indexes. Water Resources Research 27(9):2381-2399.

Resh, V. H., A. V. Brown, A. P. Covich, M. E. Gurtz, H. W. Li, G. W. Minshall, S. R. Reice, A. L. Sheldon, J. B. Wallace, and R. Wissmar. 1988. The role of disturbance in stream ecology. Journal of the North American Benthological Society 7:433-455.

Revelle, C. S., and K. E. Rosing. 2000. Defendens imperium romanum: a classical problem in military strategy. American Mathematical Monthly 107(7):585-594.

Revelle, C. S., J. Schweitzer, and S. Snyder. 1996. The maximal conditional covering problem. Infor. 34(2):77-91.

Richter, B. D., J. V. Baumgartner, J. Powell, and D. P. Braun. 1996. A method for assessing hydrologic alteration within ecosystems. Conservation Biology 10:1163-1174.

Riggs, H. C. 1972. Low Flow Investigations, Hydrologic Analysis and Interpretation. USGS Techniques of Water-Resources Investigations,

Chapter B1, Book 4. 17 pp. Washington, D.C.: U.S. Government Printing Office.

Rodriguez-Iturbe, I., and J. M. Megia. 1974. The design of rainfall networks in time and space. Water Resources Research 10(4):713-728.

Rosgen, D. 1994. A classification of natural rivers. Catena 22:169-199.

Rulik, M. 1997. Dynamics and Vertical Distribution of Particulate Organic Matter in River Bed Sediments in Groundwater/Surface Water Ecotones, Biological and Hydrological Interactions and Management Options. Cambridge: University Press.

Sampson, P. D., and P. Guttorp. 1992. Nonparametric estimation of nonstationary spatial covariance structure. Journal of the American Statistical Association 87(417):108-119.

Sankarasubramanian, A., and U. Lall. 2003. Flood quantiles in a changing climate: seasonal forecasts and causal relations. Water Resources Research 39(5):1134.

Sauer, V. B., and R. W. Meyer. 1992. Determination of error in individual discharge measurements. USGS Open-File Report 92-144. Reston, VA: U.S. Geological Survey.

Schalk, G. K., R. R. Holmes, Jr., and G. P. Johnson. 1998. Physical and chemical data on sediments deposited in the Missouri and the Mississippi River flood plains during the July through August 1993 flood. Floods in the Upper Mississippi River Basin 1993. USGS Circular 1120-I. Reston, VA: U.S. Geological Survey.

Schilling, D. D., C. S. Revelle, J. Cohon, and D. J. Elzinga. 1980. Some models for fire protection locational decisions. European Journal of Operational Research 5(1):1-7.

Schindler, J. E., and D. P. Krabbenhoft. 1998. The hyporheic zone as a source of dissolved organic carbon and carbon gases to a temperate forested stream. Biogeochemistry 43:157-174

Schlosser, I. J. 1987. A conceptual framework for fish communities in small warm water streams. Pp. 17-23 in Community and Evolutionary Ecology of North American Stream Fishes, W. J. Matthews and D. C. Heins (eds.), Norman, OK: University of Oklahoma Press.

Schumacher, P., and J. V. Zidek. 1993. Using prior information in designing intervention detection experiments. Annals of Statistics 21(1): 447-463.

Scott, M. L., G. T. Auble, and J. M. Friedman. 1997. Flood dependency of cottonwood establishment along the Missouri River, Montana, USA. Ecological Applications 7: 677–690.

Serra, D., V. Marianov, and C. S. Revelle. 1992. The maximum-capture hierarchical location problem. 1992. European Journal of Operations Research 62(3):363-371.

Shannon, C. E. 1948. A mathematical theory of communication. Bell Syst. Tech. J. 27(3, 4):379-423, 623-656.

Slack, J. R., and J. M. Landwehr. 1992. Hydro-Climatic Data Network: A U.S. Geological Survey streamflow data set for the United States for the study of climate variations, 1974-1988. U.S. Geological Survey Rep. 92-129. Reston, VA: USGS.

Slack, J. R., A. M. Lumb, and J. M. Landwehr. 1993. Hydro-Climatic Data Network (HCDN): Streamflow Data Set, 1874-1988. U.S. Geological Survey Water-Resources Investigations Report 93-4076 (CD-ROM).

Slade, R. M., Jr., T. Howard, and R. Anaya. 2001. Evaluation of the streamflow-gaging network of Texas and a proposed core network. Water Resources Investigations Report 01-4156. Austin, TX: U.S. Geological Survey. 40 pp.

Smith, L. C., and D. E. Alsdorf. 1997. Flood mapping from phase decorrelation of tandem ERS data; Ob' River, Siberia, space at the service of our environment. European Space Agency Special Publication 414: 537-539.

Southard, R. 1995. Flood volumes in the Upper Mississippi River Basin April 1 through September 30 1993. USGS Circular 1120-h. Reston, VA: U.S. Geological Survey.

Spicer, K. R., J. E. Costa, and G. Placzek. 1997. Measuring flood discharge in unstable stream channels using ground-penetrating radar. Geology 25(5):423-426.

Stanford, J. A., and J. V. Ward. 1988. The hyporheic habitat of river ecosystems. Nature 335:64-66.

Stedinger, J. R., and G. D. Tasker. 1985. Regional hydrologic analysis—ordinary weighted and generalized least squares compared. Water Resources Research 21(9):1421-1432.

Stoertz, M. W., M. L. Hughes, N. S. Wanner, and M.E. Farley. 2001. Long-term water-quality trends at a sealed, partially flooded underground mine. Environmental & Engineering Geoscience 7(1):51-65.

Subramaniam, P. 2001. Optimal Locations of Booster Stations in Water Distribution Systems. Master of Science Thesis. University of Cincinnati.

Sudau, A. 2002. Personal Communication. Bundesanstalt für Gewässerkunde, Referat Geodäsie. Federal Institute of Hydrology, Germany.

Swenson, H. A. 1962. Water research by the Geological Survey. Military Engineer Vol. 54, No. 359. Washington, D.C.: Society of American Military Engineers.

Swersey, A. J. 1994. The deployment of police, fire, and emergency medical units. Pp. 151-200 in Operations Research and the Public Sector

S. M. Pollock, M. H. Rothkopf, and A. Barnett (eds.). Amsterdam, New York: North Holland.

Swersey, A. J., and L. S. Thakur. 1995. An integer programming model for locating vehicle emissions testing stations. Management Science 41(3): 459-512.

Tasker, G. D. 1986. Generating efficient gaging plans for regional information. In Integrated Design of Hydrological Networks, Proceedings of the Budapest Symposium. IAHS Publication No. 158:269-281.

Taylor, H. E., R. C. Antweiler, T. I . Brinton, D. A. Roth, and J. .A. Moody. 1994. Major ions, nutrients, and trace elements in the Mississippi River near Thebes, Illionois, July through September 1993. USGS Circular 1120-d. Reston, VA: U.S. Geological Survey.

Team for Evaluating the Wisconsin Water-Monitoring Network. 1998. An integrated water-monitoring network for Wisconsin. Wisconsin Water Resources Institute Special Report, WRI SR 96-01. Madison, WI: University of Wisconsin-Madison. 36 pp.

Townsend, C. R. 1989. The patch dynamics concept of stream community ecology. Journal of the North American Benthological Society 8:36-50.

Townsend, C. R., and A. G. Hildrew. 1994. Species traits in relation to a habitat templet for river systems. Freshwater Biology 31:265-276.

Triska, F. J., J. H. Duff, and R. J. Avanzino. 1993. The role of water exchange between a stream channel and its hyporheic zone in nitrogen cycling at the terrestrial-aquatic interface. Hydrobiologia 251:167-184.

U. S. Geological Survey (USGS). 1998. A New Evaluation of the USGS Streamgaging Network: A Report to Congress dated November 30, 1998. Reston, VA: U.S. Geological Survey.

U. S. Geological Survey (USGS). 1999. Streamflow Information for the Next Century. Open File Report Report 99-456. Reston, VA: U.S. Geological Survey.

U.S. House Appropriations Subcommittee on Interior and Related Agencies. 1998. Report to the USGS Accompanying the Fiscal Year 1999 Appropriations Bill, H.R. 4193.

Vervier, P., J. Gibert, P. Marmonier, and M. J. Pare-Olivier. 1992. A perspective on the permeability of the surface freshwater-groundwater ecotone. Journal of the North American Benthological Society 11:93-102.

Vogel R. M., I. Wilson, and C. Daly. 1999. Regional regression models of annual streamflow for the United States. Journal of Irrigation and Drainage Engineering-ASCE 125(3):148-157.

Vörösmarty, C. J., P. Green, J. Salisbury, and R. B. Lammers. 2000. Global water resources: vulnerability from climate change and population growth. Science 289:284-288.

Wagner, B. J. 1999. Evaluating data worth for ground-water management under uncertainty. Journal of Water Resources Planning and Management 125(5):281-288
Wahl, K. L., W. O. Thomas, Jr., and R. M. Hirsch. 1995. Stream-Gaging Program of the U.S. Geological Survey. USGS Circular 1123. Reston, VA: U.S. Geological Survey.
Webb, R. H., J. C. Schmidt, G. R. Marzolf, and R. A. Valdez. 1999. The controlled flood in Grand Canyon. American Geophysical Union Monograph No. 110. 367 pp.
Winter, T. C., J. W. Harvey, O. L. Franke, and W. M. Alley. 1998. Ground water and Surface Water: A Single Resource. USGS Circular 1139. Denver, CO: U.S. Geological Survey. 79 pp.
Wörman, A., A. I. Packman, H. Johansson, and K. Jonsson. 2002. Effect of flow-induced exchange in hyporheic zones on longitudinal transport of solutes in streams and rivers. Water Resources Research 38(1):10.1029/2001WR000769.
Zidek, J. V., W. M. Sun, and N. D. Le. 2000. Designing and integrating composite networks for monitoring multivariate Gaussian pollution fields. Journal of the Royal Statistical Society Series C-Applied Statistics 49:63-79.

Appendix A
Biographical Sketches of Members of the Committee on Review of the USGS National Streamflow Information Program

David R. Maidment, *chair*, is the Ashley H. Priddy Centennial Professor of Engineering and director of the Center for Research in Water Resources at the University of Texas at Austin. He is an acknowledged leader in the application of geographic information systems (GIS) to hydrologic modeling. His current research involves the application of GIS to floodplain mapping, water quality modeling, water resources assessment, hydrologic simulation, surface water-groundwater interaction, and global hydrology. He is the coauthor of *Applied Hydrology* (McGraw-Hill, 1988) and the editor-in-chief of *Handbook of Hydrology* (McGraw-Hill, 1993). From 1992 to 1995 he was editor of the *Journal of Hydrology*, and he is currently an associate editor of that journal and of the *Journal of Hydrologic Engineering*. He received his B.S. degree in agricultural engineering from the University of Canterbury, Christchurch, New Zealand, and his M.S. and Ph.D. degrees in civil engineering from the University of Illinois at Urbana-Champaign.

A. Allen Bradley, Jr. is an associate professor of civil and environmental engineering at the University of Iowa and a research engineer at IIHR Hydroscience & Engineering. His research interests are in the areas of hydrology and hydrometeorology, including flood and drought hydrology, hydroclimate forecasting, and water resource applications of remote sensing. He received his B.S. in civil engineering from Virginia Tech, an M.S. in civil engineering from Stanford University, and a Ph.D. in civil and environmental engineering from the University of Wisconsin.

Benedykt Dziegielewski is professor of geography at Southern Illinois University at Carbondale and executive director of the International

Water Resources Association. His two main research areas are water demand management (urban water conservation planning and evaluation, water demand forecasting, modeling of water use in urban sectors) and urban drought (drought planning and management; measurement of economic, social, and environmental drought impacts). He is editor-in-chief of *Water International* and is an honorary lifetime member of the Water Conservation Committee of the American Water Works Association. He received his B.S. and M.S. in environmental engineering from Wroclaw Polytechnic University, Wroclaw, Poland, and his Ph.D. in geography and environmental engineering from Southern Illinois University.

Richard Howitt is professor of economics at the University of California-Davis. Dr. Howitt's research focuses on resource and environmental economics, quantitative methods, and econometrics. His interests include developing calibration methods based on maximum entropy estimators to model the economic structure of resource use from disaggregated physical data, including remote sensing methods, to infer the underlying economic functions. Much of his research has focused on California's water resources, including water markets in the San Joaquin Valley and the Westlands Water District. He has published in such areas as river water quality, water use, water management, and water institutions. Dr. Howitt received his Ph.D. and M.S. degrees in economics from the University of California-Davis.

N. LeRoy Poff is an associate professor in the Biology Department of Colorado State University. Dr. Poff received a B.A. in biology from Hendrix College, an M.S. in environmental sciences from Indiana University in Bloomington, and a Ph.D. in biology from Colorado State University. His primary research interests are in stream and aquatic ecology and in quantifying the responses of riverine ecosystems to natural and altered hydrologic regimes, from local to watershed to regional scales. Dr. Poff has served as a member of the Adaptive Management Forum for CALFED river restoration projects, the Scientific Review Team for the King County (Seattle, WA) Normative Flows Project, the Scientific and Technical Advisory Committee for American Rivers, and the Scientific Advisory Board of the David H. Smith Conservation Research Fellowship Program for The Nature Conservancy. He is also an Aldo Leopold Leadership Fellow of the Ecological Society of America.

Karen L. Prestegaard is an associate professor of geology at the University of Maryland. Her research interests include sediment transport and depositional processes in mountain gravel-bed streams; mechanisms of streamflow generation and their variations with watershed scale, geology, and land use; hydrologic behavior of frozen ground; hydrologic consequences of climate change; and hydrology of coastal and riparian wetlands. She was a member of the National Research Council (NRC) Committee for Yucca Mountain Peer Review: Surface Characteristics, Preclosure Hydrology, and Erosion. She received her B.A. in geology from the University of Wisconsin-Madison, and her M.S. and Ph.D. in geology from the University of California, Berkeley.

Stuart S. Schwartz is director of the Center for Environmental Science, Technology, and Policy at Cleveland State University (CSU). Before joining CSU, Dr. Schwartz served as associate director of the Water Resources Research Institute of the University of North Carolina. Previously, Dr. Schwartz served as an associate hydrologic engineer at the Hydrologic Research Center in San Diego, California, and directed the Section for Cooperative Water Supply Operations on the Potomac at the Interstate Commission on the Potomac River Basin. Dr. Schwartz's research and professional interests are in the application of probabilistic hydrologic forecasting and multiobjective decision making in risk-based water resources management, watershed management, and water supply systems operations. He received his B.S. and M.S. in biology-geology from the University of Rochester and a Ph.D. in systems analysis from the Johns Hopkins University.

Donald I. Siegel is a professor of geology at Syracuse University, where he teaches graduate courses in hydrogeology and aqueous geochemistry. He holds B.S. and M.S. degrees in geology from the University of Rhode Island and Pennsylvania State University, respectively, and a Ph.D. in hydrogeology from the University of Minnesota. His research interests are in solute transport at both local and regional scales, wetland-groundwater interaction, and paleohydrogeology. He was a member of two NRC committees: Committee on Techniques for Assessing Ground Water Vulnerability and Committee on Wetlands Characterization.

Mary W. Stoertz is an associate professor of hydrogeology at Ohio University, Department of Geological Sciences. Her area of specialty is stream restoration, especially restoration of channelized rivers and streams polluted by acid mine drainage. She founded the Appalachian Watershed Research Group at Ohio University, which has the mission of restoring desired functions of watersheds subject to mining, sedimentation, and flooding. She directs the multidisciplinary research arms of the Monday Creek

Restoration Project and the Raccoon Creek Improvement Committee. Dr. Stoertz received her B.S. in geology from the University of Washington and her M.S. and Ph.D. in hydrogeology (with a minor in civil and environmental engineering) from the University of Wisconsin-Madison.

David G. Tarboton is professor, Utah Water Research Laboratory and Department of Civil and Environmental Engineering, Utah State University. His research interests are in spatially distributed hydrologic modeling, applying digital elevation data and GIS in hydrology, stochastic hydrology using nonparametric techniques, snow hydrology, geomorphology, landform evolution and channel networks, and terrain stability mapping and stream sediment inputs. He is a member of the American Geophysical Union, American Society of Civil Engineers, and American Water Resources Association and is a registered professional engineer (Utah). Dr. Tarboton received his B.S. in civil engineering from the University of Natal in Durban, South Africa, in 1981, and an M.S. and Sc.D. in civil engineering from the Massachusetts Institute of Technology in 1987 and 1990, respectively.

Kay D. Thompson is a consultant. In her research she investigates properties of subsurface materials for groundwater studies, develops methods for subsurface characterization, assesses the risks of hydrologic dam failure, and consults on minimizing environmental impacts during development. Dr. Thompson received a B.S. in civil engineering and operations research in 1987 from Princeton University, an M.S. in 1990 from Cornell University, and a Ph.D. in 1994 in civil and environmental engineering from the Massachusetts Institute of Technology. Dr. Thompson was formerly an assistant professor at Washington University, Department of Civil Engineering.